BEFORE YOU PAY ANOTHER UTILITY BILL— BEFORE YOU MAKE A SINGLE HOME IMPROVEMENT READ THIS BOOK!

Any average homeowner or small businessman who has little extra time and less extra money can start saving on both with the valuable information in these pages. Here is a DO-IT-NOW-PLAN that trims the cost of energy right away—

Learn how to save $20 with the flick of a switch . . . Learn how the government will help you lower the price tag on your energy consumption . . .
and much more in—

THE HOME ENERGY GUIDE

THE HOME ENERGY GUIDE

How to Cut Your Utility Bills

John Rothchild
and
Frank F. Tenney, Jr.

BALLANTINE BOOKS • NEW YORK

Copyright © 1978 by John Rothchild and Frank F. Tenney, Jr.

All rights reserved under International and Pan-American Copyright Conventions. Published in the United States by Ballantine Books, a division of Random House, Inc., New York, and simultaneously in Canada by Ballantine Books of Canada, Ltd., Toronto, Canada.

Library of Congress Catalog Card Number: 77-18675

ISBN 0-345-27677-9

Manufactured in the United States of America

First Ballantine Books Edition: February 1978

Acknowledgments

Grateful acknowledgment is made to the following for permission to reprint previously published material:

Center for Science in the Public Interest: For a listing of what local communities are doing to conserve energy.

Energy Research and Development Administration, Office of Public Relations: For a reproduction of charts from "An Economic Analysis of Solar Water and Space Heating," November 1976 and "Solar Energy for Space Heating and Hot Water."

Environmental Action Foundation: For a listing of various conservation programs offered by utility companies from *Taking Charge*. Copyright © 1976.

Dr. Jay McGrew, applied science and engineering: For information on insolation.

Mutual Redevelopment Houses, Inc.: For an analysis of energy conservation in a cooperative apartment building.

Popular Science Magazine: For a chart on heat pumps and information from articles which appeared in the following issues: September 1975; October 1975; July 1976; September 1976; and December 1976.

Grateful acknowledgment is made to the following for permission to reprint photographs, diagrams, and reproductions:

Camberlain Manufacturing Company: A photograph of solar collector panels at NASA's Langley Research Center, Hampton, Virginia.

Inframetrics, Inc.: A photograph of a portable infrared scanner system and a reproduction of a thermograph.

Libbey-Owens-Ford Company: A diagram of a typical flat-plate solar collector system and a photograph of a solar home.

Malleable Iron Range Company: A diagram of an add-a-furnace.

Vermont Casting Company: A reproduction of a defiant parlor stove.

Contents

1. **Introduction**   1

2. **The Energy Budget: Staking Out the Meter**   11
   A Family Energy Manager • The Myths of Energy • The Energy Audit • Finding the Price Tag • Marking Your Progress • Reading the Appliance Labels • Calculating the Costs of Your Four Energy Systems • Small Appliances • Reading the Meter

3. **Shedding the Load: How to Save Money without Spending Any**   44
   Hot Water • Heating • Cooling or Air Conditioning • Cooking and the Kitchen • Refrigerator and Freezer • Lighting • Entertainment (and Miscellaneous Items) • Special: Saving Money on Water

4. **Energy Gadgets: Where the Payback Exceeds the Price**   85
   Hot Water • Heating • Cooling or Air Conditioning • Cooking and the Kitchen • Refrigerator and Freezer • Lighting • Entertainment (and Miscellaneous Items) • Special: Saving Water • Additional Energy-Saving Ideas

5. **The Hidden Dividend: How to Buy Large and Small Appliances and Systems**   108
   Understanding Life-Cycle Costing • Air Conditioners: The Money's in the EER • Refrigerators, Freezers, and Deep Freezers • Heaters and Furnaces • Stoves and Ovens • Hot-Water Heaters • Washing Machines • Dryers • Dishwashers • Televisions • Waterless Toilets • Further Information

6. **Tightening Up the House: An Extra Blanket Doesn't Always Pay**   131
   Introduction • Insulation • Plugging the Leaks • Installing Storm Windows • Installing Storm Doors • A Plan of Action

7. Working Up to Solar — 149
The Solar Greenhouse: A Simpler Collector • The Heat-Recovery Unit: What Can the Sun Do That an Air Conditioner Can't? • Windmills: An Overblown Prospect

8. How to Buy Solar: Or Do You Need It? — 155
The Solar Water Heater • The Solar Space Heater • Swimming-Pool Solar Heaters • Solar Air Conditioners • A Solar Catalogue and Suggestions for Further Reading • Solar-Product Manufacturers

9. Energy Survival: What to Do Before the Power Fails — 180
The Survival Room

10. The Apartment Connection: Ways to Save Energy in Big Buildings — 186
Investigating Your Energy Rights as a Renter • Banding Together for Fun and Profit—In a Master-Metered Building or Co-Op

11. The Public Goodies: Utilities, Banks, and the Government — 196
Getting the Most Out of a Utility Company • Banks and Other Loan Sources • Help from the State and Federal Governments: Energy Saving and Taxes • Grants

12. "Turning Down" the Town: They Did It in Seattle — 210
The Successes of Seattle and Los Angeles • Should Energy Conservation Be Made Mandatory? • The Hartford Plan and Local Energy Conservation • Unplugging Your Town • For Further Information . . .

13. Is There an Energy-Saving House in Your Future? — 224
The Old-Timer Factor: Traditional Local Knowledge • An Energy-Efficient Residence, or EER • A Traditional House with Some Energy-Saving Features • Adding a Room to an Existing House

14. Conclusion: What If Everybody . . . ? — 244

Acknowledgments — 247

SECTION 1

Introduction

This book is for people who enjoy their comforts and conveniences and would like to maintain them—as much as possible—in the face of rising electricity and fuel-oil prices. It is for people who want to survive the energy crisis without moving to a cave in New Mexico or cooking on recycled newspapers. It is for people who may be somewhat concerned about how much oil might be left under the subsurface of the earth, but who are more concerned with the problem of paying for it, delivered. If you are one of these people, you are probably facing your own *monthly* Energy Crisis—in the form of the utility bill itself. Those bills once nibbled at the paycheck, but now they gobble like a second mortgage. You can't do much to reduce the worldwide dependence on oil, but you stand a good chance of reducing your own energy costs to a tolerable level. This is one national crisis you can solve on your own.

The only question is, what to do.

There is plenty of advice around, but most of it takes one of two unsatisfactory approaches. The first approach is the passionate espousal of the obvious: You have probably already been reminded that leaving the light bulbs burning wastes electricity, that opening the front door during the winter allows heat to escape, and so on. Such prescriptions wouldn't *seem* to have any more economic benefit to you than picking up pennies off the street; many of them *don't*, unless they are coordinated into a systematic plan of action. We will avoid the scatter-gun tactic of offering obvious and disconnected suggestions.

The second approach takes energy conservation way off in another direction, toward the far reaches of solar physics and thermal theory—advice that is lost in the confusing morass of terms like "joules" and "convection forces" and "BTUs" and "degree-days," charts that look like diagrams of busy freeways, and do-it-yourself plans that cannot be followed without a working knowledge of physics, carpentry, roof repair, electrical engineering, celestial navigation, solar dynamics, and how to survive in the woods. Such energy-saving advice and material is sometimes very inspiring, and often useful to the people who understand it, but it is light-years away from the reality of most homeowners (or apartment dwellers)—people who don't have time or ability to build concentrating collectors or self-propelling energy systems, and still aren't sure of the more basic questions, like "Does it really save money to turn off the hot-water heater overnight?" or "Could it possibly make a difference to toast bread in the oven instead of in the toaster?"

Between these two approaches—the trivial solutions that don't seem to offer a real chance to save money, and the complicated solutions that don't seem directed at the average person—there doesn't *appear* to be a middle ground. This book is an attempt to find one.

Our attempt is hopefully enhanced by the fact that two of us have collaborated in the project of producing this book.

Frank has an engineering background, and he understands the technical details that leave most people glassy-eyed and confused. He has built his own solar heater, and it works.

John is a journalist. He tried to build a solar heater, but the parts still sit on the ground outside his house: the wrong-size copper pipes, the corroded aluminum, and the decaying wood—as a kind of monument of a do-it-yourself fiasco.

Frank knows what he is talking about, and many of the ideas in this book have been tested and applied in his Florida home. He continues to use all the machinery that his neighbors use, but his energy-saving approach results in a $50-a-month summer electric bill, instead of the usual $180.

John knows the limitations of non-technical people, because he is one of them. He has already been led astray by some of the how-to information drifting around, and he hopes to spare others the same fate.

We have tried to discard overly technical or fanciful material.

In the course of researching this project—from already-published information, interviews with backyard inventors, and unbelievable stacks of material provided by the federal government—we are convinced that *you can cut* from *20 to 50 percent* off your heating, cooling, and other utility bills. You can do it, and still live comfortably, without unplugging all the appliances that have been developed in the last fifty years to make domestic work easier. You may even be able to save that first 25 percent, or around $200 a year, without spending any money on insulation or gadgetry. But it will require a little effort. This isn't one of those miracle diets that claim you can eat anything you want and still lose weight. The effort has to be consistent.

Many people attempt energy conservation in periodic bursts of zeal. The bursts usually occur around the time they receive another utility whopper, when the bill payer is likely to march into the living room, flip off a light in a sanctimonious flourish, and give the children a short lecture on applying the virtues of Franklin to the light bulbs of Edison: "A kilowatt saved is a penny earned!" The trouble with such efforts is that they are not repeated until the receipt of the next utility bill, which, being as large as the preceding one, tends to give energy saving a bad name. This kind of disjointed approach, based on one tip this week, another tip next week, has not proven effective.

What is needed, especially for the 25-percent reduction that can be achieved at no expense to the homeowner, is a plan.

Corporations already have energy plans, which they have been following for three or four years, with remarkable success. These corporate procedures have names like "Total Energy Management" or "Total Energy Budget"; but the general idea is always the same. Together with a lot of government and private research help, businesses discovered that telling the workers to turn off faucets and close doors behind them wasn't going to conserve many kilowatts, and that what was needed was a complete understanding of how energy was used and misused through an entire plant, from top to bottom. Energy audits were devised, and corporations could see how work patterns could be changed, and building structure modified, to provide an easy 20- to 30-percent drop in utility bills. The General Services Administration achieved 30 percent in many public buildings. Corporations were so happy with the plans that they have been codified in a series of government manuals. No such manuals have been produced for the residential energy user, but we think our plans will work for you the same way they have worked for corporate management.

There are ways to use this book without following the energy plan, step-by-step, through the various sections. You can skip to the no-cost suggestions in Section 3 and find a series of quick changes—reducing the water heater thermostat setting, or unplugging the sump heater on the central air conditioner—that will save you more than a year's worth of turning off lights. You can read about the gadgets in Section 4: relatively inexpensive items that pay back in a year or two, and in some cases return tremendous dividends. Or you could go right to the solar heaters or the insulation sections, to decide how to make your major energy-saving investments. But to maximize your savings, and to minimize the amount of actual money you must put up to achieve them, we suggest that you follow the plan from beginning to end.

The purpose of the plan is to teach you how to make an energy budget.

Most people already budget their food and their entertainment and their phone bills, and while the utility bill was once too trivial for such scrutiny, it has crept up in importance, and the skills to understand it have not been generally developed. An energy budget can save at least as much as a food budget, with about the same amount of effort. The problem is how to make one. The utility bill remains a mystery, and its timing seems calculated to produce consumer indifference. In fact, it arrives about two roast turkeys and three laundry loads into the *next* billing period. The final result always seems disconnected from the activities that produced it.

The first part of the budget involves a quick energy audit. The idea is that you will never know how much you are saving until you know how much you are spending. Most people have only hazy ideas about what each of their appliances costs them to operate, especially if the heating system and the cooling system and the kitchen and laundry are all thrown in on the same utility bill. This haziness not only makes it hard to know *where* to reduce the utility bill, it leaves people open to exaggerated claims of insulation companies or solar heater companies as to the potential savings on their various products. You will never know whether solar heaters or insulation are a good investment until you figure out how much it costs you to heat your water or heat your home.

The energy audit can be effected, courtesy of the utility company's own meter, which is usually installed in your basement or backyard, or in your apartment kitchen or basement. That meter is one of the best energy-saving devices you can find; in fact, one conservation experiment is placing an easy-to-read meter in the kitchen of the household, so people will be more inclined to use it. By learning to read your meter, you can tell exactly which of your appliances cost you the most money, how the various activities of the day combine to create a high utility bill, how the weather conditions are affecting your fuel consumption, and what you can do to use the weather itself as an energy-saving tool. You will also be able to decide, on a day-to-day basis, whether the energy-saving suggestions you have adopted are working, or whether you should abandon them and try something else. Most important, the meter will give you a method of "weighing" your daily energy consumption, which is as important to kilowatt reducing as a scale is to fat reducing. (The meter is much more useful than the lists published by utilities companies concerning appliances and their estimated energy consumption. Nevertheless, for people who don't have a meter, or don't want to bother to read it, we will provide a way to estimate the cost of running the various appliances, too, in Section 3.)

All the corporations began their energy programs with audits, and once you have yours, then you can begin to save money in a systematic way. Some people would suggest that you go right out and buy all the insulation or energy gadgets that you can afford, but our advice is to wait and see what you can save without buying anything. You can use the knowledge you gain in the process in order to make a *better choice* of equipment, and you can even use the money you save now to make purchases later. That way, your energy-conservation plan will at least partially finance itself!

We wouldn't make such a suggestion if we didn't believe that families can save significant amounts of money by altering their energy habits only slightly, and by using their houses and appliances more intelligently. Even many of the experts, until recently, did not think that personal actions and habits had much effect on fuel or electric bills. The "hardware people," as they are called, went around studying air conditioners and furnaces, and forgot to look at the people who used them.

One group of researchers was sent to analyze a residential community at Twin Rivers, New Jersey. They made the surprising discovery that similar families, living in identical townhouses and with all the same appliances, did not use the same amount of electricity or gas. In fact, one family at Twin Rivers might use 50 percent more electricity or gas than its neighbors. The researchers carefully studied the structure of the houses, looking for cracks or leaks or something to explain the discrepancies. But the answer couldn't be found in the walls of houses.

The hardware people finally realized that the difference in energy consumption had more to do with personal habits than anything else, and they called in the psychologists. If two families could live at the same level of comfort, and yet one could use 50 percent less fuel, that meant that the way people handled their appliances and their furnaces, the way they controlled their living space, could have a tremendous effect on utility bills. And, it meant that people really could save a lot of money *without buying expensive products* and *without giving up their conveniences*. A new and interesting research program was begun at Princeton, a program that focuses on people as well as on buildings and thermal dynamics.

The structure of this book owes a lot to the discoveries made at Princeton. Since the studies convinced us that people could do a lot on their own, we decided to emphasize people first and hardware second. If you are a high user by instinct, we hope the plan contained here will turn you into a low user—and cut your bills in half. If you are a low user already, we won't be able to save you as much, but we are still certain we can save you a significant amount. Our belief is that you can cut from 5 to 10 percent off energy bills just through general awareness, and another 10 to 15 percent through the no-cost suggestions offered in the third section of the book that should give you the original 25 percent.

By that time—the end of Section 3—you will be ready to consider insulation, solar heaters, and other products that will save you the final 25 to 30 percent. That's not to say that insulation wouldn't have an immediate impact on how much money you spend for fuel. As a matter of fact, if your house has no ceiling insulation, we advise you to buy some immediately, because the economic returns will be tremendous. But if your house is already partly insulated, and you aren't sure what to do next, then we suggest starting with the things that can be done for free.

The no-cost suggestions of Section 3 are presented in an appliance-to-appliance tour of a regular home. They are lucrative suggestions, if they are orchestrated in the proper way. That's where the plan comes in. We emphasize small adjustments, like the hot water heater thermostat, that have immediate paybacks. We also discuss cumulative adjustments, as in cooking, in which the operation of one appliance can have an effect on the cost of running another. (It is said that for every $2 you save by limiting other appliance usage in the summer, you save an additional $1 in air-conditioning bills.) And finally, we recommend spatial adjustments, such as reducing the number of rooms to be heated or cooled, the use of insulative blinds and materials inside the house, and even the possible addition of a survival room (that can be heated at minimal expense). The survival room will help people get through crisis situations like blackouts and fuel shortages. But it can also be used in more routine situations, such as the extremely cold weeks during a winter, when fuel bills threaten to devastate the budget.

Some of our suggestions do require a little work. But "energy chores," as we call them, can easily be done by children, or shared by all members of the family. We have all gained a lot of freedom because of our work-saving devices. When people are convinced of the benefits, they can be convinced easily to give back a few minutes of their time so that those appliances work on less fuel. If you can begin to achieve the 15- to 25-percent reduction in electricity and fuel-oil costs, you are talking about $100 to $200 a year on actions that cost nothing. Energy-conservation dividends are particularly attractive. Unlike paychecks, they are not taxed; so for every $1 you save, you are actually getting back $1.20. They are also adjusted for inflation. Every time fuel prices go up, the dividends from energy conservation go up in proportion. Starting a conservation program is like owning tax-free, guaranteed bonds, adjusted to the cost of living. A $200-a-year dividend is like owning more than two thousand dollars' worth of stock in the utility company. Electric and gas prices and fuel-oil have already gone up precipitously in recent years. When they continue to go up, the marginal savings *now* may turn into crucial savings *later*.

From Section 4 onward, the chapter headings speak

for themselves. We have tried to produce a complete energy savers' guide: the up-to-date information on how to buy appliances, how to insulate, how to save energy in apartment buildings, what to do if you rent your apartment, how to find tax breaks and government goodies, how to survive blackouts, what to look for in the future—with constant attention to the economics of the situation. Surprises come in every section. Buying one model of refrigerator or air conditioner over another, we discover, can make a difference of hundreds of dollars in energy costs over the lifetime of the appliance. Learning how to buy appliances will, in fact, save you much more money than learning how to use them more judiciously. Insulating one part of the house may be a lucrative proposition, while insulating another part may not. Weather-stripping and caulking can be crucial—but not necessarily in the places that you hear about. Solar hot-water heaters may be good investments, but the heat-recovery unit—a much cheaper item—may do the same job and do it better.

Since we are not an appliance testing service, we cannot guarantee the performance of any of the products we discuss in the book. But we have tried to present items that have proven themselves in the marketplace.

A couple of big projects, coupled with whatever no-cost activities you have undertaken, can reduce your electric and gas and fuel-oil bills by as much as 50 percent. Fifty percent is the final goal for the plan. By choosing the proper items for your situation, and by budgeting your energy, you should be able to reach the goal within a couple of years. And if $200 was a good dividend for the no-cost suggestions, $500 a year should be a worthwhile reward for the whole energy plan. To get a $500-a-year tax-free dividend anyplace else, you would have to own more than eight or nine thousand dollars' worth of stocks and bonds.

If you can begin to reduce your family energy consumption now, you may enjoy advantages that go beyond the money you have saved. You will be prepared for changes in utility policies, such as off-peak pricing that penalize people heavily for using appliances during daytime hours. You will be able to cope with the major increases in fuel and electric costs predicted for the next fifteen years, when energy saving will become more of a necessity. You will be prepared, if you design a survival room, for blackouts, brownouts, and fuel shortages. And the government is talking about mandatory energy rationing—if voluntary efforts do not work. If you have begun the process of controlling your energy budget, such prospects will not be as difficult to face. If you can cut energy bills while it is still a matter of choice, you will be ready for the time when it becomes a matter of necessity.

> Frank Tenney
> John Rothchild
> December 1977

SECTION 2

The Energy Budget: Staking Out the Meter

A Family Energy Manager: You've Got to Have a Goal

Corporations that have undertaken successful energy-saving programs have discovered that *one person* has to be ultimately responsible.

Little money can be saved in a chaotic situation in which people attack the utility bill through sporadic gestures, like turning off each other's lights or jiggling and unjiggling the thermostats. A coordinated effort, with one person doing most of the budgeting and cheerleading, is what has brought substantial reductions to the energy costs of large businesses. Such businesses have recognized the economic importance of energy programs by bestowing the title of Energy Manager on the person they choose to direct these. Your household might consider appointing an energy manager.

Someone in your family probably spends a lot of time penciling-in the food budget and keeping track of other major purchases. The time it takes to control the energy bills can be equally rewarding. Energy manager might be a good job for the most even-tempered nitpicker in your family. Or it could be the ideal part-time occupation of an older child, who can learn something about how appliances work and save you money at the same time. For imaginative teenagers, energy managing is a promising new prospect—

lucrative, useful, and less strenuous than mowing the neighbor's lawn. Later, as *neighborhood* energy consultants, they can show other families how to cut utility bills, then take some percentage of the amounts saved.

The first task for an energy manager is to consider some form of motivation.

People don't change their eating routine *just* because they have read an article about the benefits of being thin. The same is true of the energy routine. The amount of electricity, gas, and fuel oil that we use is determined by habits that will not be altered without some regular reminders and some definite incentives. The citizens of Los Angeles got an incentive in 1973, when the city advised them they would have to make drastic and immediate attempts to conserve power, or else the utility would run out of oil to operate its generators. Residential customers responded by an immediate 18-percent reduction in energy use (see Section 12 for more details). Every homeowner would be happy to get the annual return of $200 or more that such a reduction would produce. But most homeowners do not know how to motivate themselves to make the effort.

Researchers at Princeton University have been wrestling with this problem of motivation, and have reached interesting conclusions. They found that people in their test community, Twin Rivers, New Jersey, could cut impressive amounts off their electric bills if they received one reminder a day. Convinced, by their early findings, that people can save as much through desire as they can through devices, the Princeton researchers refined their motivational techniques. They recruited various homeowners for energy-saving experiments and divided them into different groups. One group was given a very hard goal: a 20-percent reduction in energy use. Another group got an easy goal: a 2-percent reduction. Some families within each group were given daily feedback on their progress, others were not. The results are instructive. The easy-goal, no-feedback group actually used *more* energy than a control group that wasn't told to do anything.

The group that was given a difficult goal but got no feedback managed to save only a small amount of energy, 1.8 percent over the control. In fact, the only group that registered savings of any significance was the one that got a difficult goal and also got feedback; that group used 13.4 percent less energy than the control.

The 13.4-percent reduction was achieved without any new gadgets, extra insulation, or solar devices, and without any specific plan for how to save energy. The fact that some families got 13.4 percent through the mere power of suggestion gives cause for optimism. But the results also suggest that you need two things to pursue a successful energy program: a specific, and fairly substantial, goal; and some kind of daily progress report.

Since you probably can't get Princeton University to call you on the phone every day, *you* will have to invent your own goal and deliver your own feedback. A tentative first goal might be 15 percent—an amount that can be achieved on awareness alone. As you get farther into the no-cost section, you can raise the goal to 25 percent and, by the end of the plan, you should be trying to reach the 50 percent. You can get the first 15 percent by eliminating inefficient and wasteful practices. Even if you have already been told that too much traffic in and out of the refrigerator can cost several dollars a year in added electric bills, you probably aren't compelled to save that small amount. But when you have the overall goal in mind, you may take such contributing factors more seriously.

If the 15-percent goal is not incentive enough in itself, you can also institute a reward or punishment system, depending upon your charitable or punitive inclinations.

A reward system would provide rebates on money actually saved through energy conservation. The dividend could go either to the energy manager, or to the entire family—which could then congratulate itself with a dinner or a movie where Reddy Kilowatt picks up the tab.

Some families have even connected the energy reward directly to the appliance that saved them money. One Virginia household, for instance, used the $125 a month they saved by turning off the air conditioner to join a club that had a swimming pool. The swimming pool kept them cool on the hottest days, helped them forget the air conditioner, and provided entertainment as well. Similarly, a family in the North that manages to reduce the heating bill might reward itself with a weekend trip to Nassau during the next cold spell.

If you prefer negative reinforcement, you can set up an energy fine system, making people pay for leaving the lights or television sets on, for lingering in the shower, or for boiling too much water. The fine money can be added to the money saved through conservation, and the total amount can be used to buy caulking, or water-saving showerheads, or other equipment that will return additional dividends. Some people have developed cumulative energy plans—first reducing their utility bills through no-cost methods, then using that "profit" to buy insulation and improvements that will further reduce the bill, and finally putting that "profit" back into solar-heating equipment; so that their conservation effort is financed primarily by money they would have spent on electricity and fuel oil in the first place.

On a neighborhood scale, it is possible to devise a contest, giving a prize to the family that saves the most electricity or fuel oil, a kind of inverse "Live Better Electrically" award. Most local newspapers would be happy to support such efforts with a little publicity.

Whatever motivation you choose, the specific goal is most important. Fifteen percent is a good place to start.

The Myths of Energy

Your early attempts to save energy may be thwarted by some of the misconceptions, or myths, that have been attached to energy use. The Greeks invented gods to explain the puzzle of the cosmos, and some of us have equally bizarre and private concepts about a subject that continues to baffle the average homeowner: how the appliances work. Nothing is further from the intellectual grasp of some of the most intelligent Americans than the inner mystery of the air conditioner, the refrigerator, or the light bulb. So, in this infancy of energy-saving times, it is not surprising that myths have developed—myths that are usually calculated to relieve people of even the slightest amount of human exertion.

Picture the situation: You are leaving the house to go to a movie, and you look across the room to count four light bulbs still burning in the table lamps. Thirty steps would be required to turn them off, but somebody has already reached you with Energy Myth #1, which says, "It takes more energy to start a light bulb than to keep it burning." So you leave the lights on. By avoiding the thirty steps, you have saved a little time, but you have not saved money. All the energy myths recorded below result in larger electric or fuel bills:

(1) *Myth.* A light bulb is like a car engine. When you turn on a light switch, a gang of little particles have to push into the filament to get it burning. It takes more power to start a light bulb than to leave it running. A corollary to this myth says it is better to leave the television set running, too.

Reality. It takes a minuscule amount of extra energy to start a light bulb or television set. The amount is so slight that you are always better off to turn it off, even for a few minutes. *Fluorescent* tubes do wear out more quickly with frequent switching on and off. But you save money by turning off incandescent bulbs and television sets whenever you can.

(2) *Myth.* A dimmer switch does not save electricity. When the dimmer is turned down low, less power is being consumed.

Reality. Most dimmers used to be rheostats, which merely absorb part of the electricity that would have gone into the light bulb and convert it into heat. Dimmers are now commonly made from solid-state components, and they do save on electricity when they are turned down.

(3) *Myth.* The way to heat a room fast is to turn the thermostat way up—say, to 90°. The furnace then works faster than it would if the thermostat were set at the desired final temperature—say, at 72°. This same myth applies to air conditioning: the house cools faster if the thermostat is set very low-temperature.

Reality. Overshooting a thermostat setting does not make a furnace or air conditioner work any faster. It saves money to set the thermostat dial exactly where you want it to remain. Then the room will never get hotter (or cooler) than necessary—causing discomfort and wasting fuel.

(4) *Myth.* The fireplace gives a big help to the regular heater. It may be a struggle to cut logs, but the extra work will save you money on fuel bills.

Reality. Looking at a fire may make you *feel* warmer, but conventional fireplaces are likely to pull more heated air up the chimney than they add to the living area. A fireplace robs your regular heater or furnace of the warm air it has generated, and extracts a double price: for the cost of wood and for the wasted heat that must be replaced by the furnace. There are ways to remedy this situation, discussed in Section 4 of this book.

(5) *Myth.* An air conditioner, like a light bulb, should be left running. It takes a surge of extra electricity to "warm up" an air conditioner. It also takes less energy to maintain a desired temperature than to *re*cool a room that has been allowed to heat up. If you are going outside for an hour, leave your air conditioners on.

Reality. The amount of energy it takes to recool a room depends on variable factors such as the size of the room, the quality of insulation, and the outside temperature. If you have a room air conditioner that does not operate on a thermostat, it is always better to turn it off, even if you are just walking down the street to go shopping. With *central* air conditioning, it will save money to turn the thermostat to a higher temperature if you leave the house or apartment for more than thirty minutes, and to turn it off if you expect to be gone longer than an hour.

(6) *Myth*. It is better to set your hot-water heater at a very high temperature. That way, less hot water will have to be mixed with cold water to produce the proper temperature at the faucet.

Reality. Much of the heat that goes to raising the temperature of the water is later lost through the walls of the water tank. The higher the water temperature inside the tank, the greater the heat losses, and the more money you pay to maintain your hot-water system.

(7) *Myth*. You save money by leaving your furnace thermostat set at the same temperature night and day. If you let the house cool down at night, the furnace will have to work extra hard to warm it up again in the morning. The extra fuel it takes to heat the house up will exceed the amount saved when the house is cooled down overnight.

Reality. It costs a lot of money to maintain the daytime thermostat setting during the colder night hours. The extra work that the furnace must do in the morning is more than offset by the fuel savings you get from the night setback. (More details are given in Section 3.)

(8) *Myth*. You save water by taking a shower instead of a bath.

Reality. Showers save on water, if the showers are *short*. A bath requires about thirty gallons of water. A standard showerhead can put out eight gallons of water per minute. A four-minute shower, therefore, may already have consumed more water than a bath; and some people are inclined to stand under the steam for a lot longer than that. Unless you view the shower as a strictly hygienic rather than a pleasure-giving device, you will probably save water by taking more baths. (We are mainly concerned about showers and/or baths because of the cost of electricity or gas to heat the water.)

The Energy Audit

Once the goal has been chosen, the motivation established, and the myths dispelled, it is time to perform an energy audit of the household. Corporations do it with the aid of a consulting firm, but you can get by with a more informal, self-help version.

For the purpose of the audit, think of your house or apartment as a group of energy systems, all working together to produce heat, light, and high utility bills. These systems are easy to identify. There's (1) the hot-water system necessary for washing clothes, dishes, and people; (2) the heating system; (3) the cooling or air-conditioning system; (4) and the energy systems for cooking and the kitchen, lighting; and (5) the systems for entertainment, and miscellaneous items.

Looking at home energy in terms of the four systems will take energy saving away from the scattergun approach. You won't save much money by applying tips from one appliance to another. But if you can take on an entire system at once—if, for instance, you attack the hot-water bill at the dishwasher, the water heater, the washing machine, and the bathroom simultaneously—the results should be meaningful and noticeable.

By learning how much each system contributes to your total energy bill, you will know the easiest and best places to start saving money. The situation varies so much from home to home that it is best for you to do your own figuring. Most published energy estimates, for instance, assume that about 20 percent of your utility bill goes to heating water. In cold climates, your hot water may even be more expensive than that, and in warm places like Davis, California, hot water comprises only 10 percent of the utility costs in some households. By doing your own energy audit, you can, also, rank your energy systems in a rough order of economic importance. If we had to make a generalized guess, we would rank them like this:

(1) *Hot water.* You might spend more money to

The Energy Budget: Staking Out the Meter   19

heat or cool the house, but hot-water systems offer the greatest possibilities for savings for the least amount of trouble or investment.

For purposes of the audit, you need only be concerned with the quantity of hot water your family requires. Estimate how many loads of wash your family goes through each week, and how many hours of drying time are required for the weekly pile of laundry. (We include the dryer as part of the hot-water system, since it is integrally connected to the washer.) Also, try to calculate how many baths or showers your family takes each week. Are they short showers or long ones? Finally, figure out how many loads of dishes are cleaned in the dishwasher each week, or, if you do dishes by hand, the number of sinkfuls of water you use each week.

Since a bath requires thirty gallons, a shower takes an average of eight to ten gallons per minute, a load of dishes in the dishwasher uses fourteen gallons, and a load of wash takes about thirty-five gallons, you can estimate, by these figures, the total number of gallons of heated water your family uses each week by filling in the blanks in the "Energy Audit" plan on page 20.

(2) *Heating.* It is hard to provide a way to estimate fuel costs of home heating, because those costs depend on so many factors. If you don't already keep track of heating bills, we'll show you a way to figure them out later in this section. If you already know the amount, fill it in in the blank on the energy-audit plan.

(3) *Cooling or air conditioning.* Same as for heating.

(4) *Cooking and the kitchen.* The percentage of energy that goes into cooking food used to be much higher than it is now, because people used to cook more than they do now. (Kilowatt charts produced by utilities don't always reflect the decreasing importance of the oven and range as energy users.) Meanwhile, the energy costs of keeping the food *cold* have risen considerably, especially since the second refrigerator and the basement freezer have become common household items. How many hot meals does your family

require each week? How many refrigerators or freezers do you operate?

(5) *Lighting.* Lights don't always offer much opportunity to save money, but if you use a lot of them, or if you keep a few lights burning all the time, they can cost you from $20 to $50 a year in extra electric bills. Make a rough estimate of the total hours of lighting your house requires each day. If you keep five lights burning for an average of ten hours apiece, that's fifty light hours per day. Pay special attention to the continuously burning lights, the multiple fixtures, and the outside decorative or protective lights; most of the money can be saved in these three places.

(5a) *Entertainment (and miscellaneous items).* Swimming-pool pumps and heaters are major fuel consumers. Do you operate a swimming-pool pump or heater, Jacuzzi or sauna bath, or any other recreational device that might require a lot of energy to run? Stereo systems may draw as much electricity as black-and-white TVs, but unless you are a constant listener, the energy money it takes to run them won't add up to much. Since the television set is the major form of entertainment for most families, and color TVs draw significant amounts of power, we concentrate on these. How much television does your family watch each week? Is the TV left on when nobody is looking at it, as a kind of supplementary lighting system for the living room or den?

Once you have completed this cursory audit of the five systems, you can fill in the results in the spaces below. The estimates will be very helpful as you continue through the energy-saving plan.

The Energy Audit

(1) Hot Water
    Number of baths times 30 gallons = _____ gallons
    Number of dishwashing loads times 14 gallons = _____ gallons
    Number of laundry loads times 35 gallons = _____ gallons

Number of showers times minutes per shower
times 10 gallons = _____ gallons
_____ total
gallons
of heated
water per week

    Clothes Dryer
      Number of loads times drying
      time = Total drying time per week   _____
(2) Heating
    Annual cost (if you know it)   _____
(3) Cooling or Air Conditioning
    Annual cost   _____
(4) Cooking and the Kitchen
    Number of hot meals per week   _____
    Number and size of refrigerators   _____
    Number and size of freezers   _____
(5) Lighting
    Number of light hours per week   _____
(5a) Entertainment (and Miscellaneous Items)
    Number of hours TV set (or sets) is
    turned on per week   _____
    Other major energy-consuming devices
    (things like sprinkler pumps, swimming-
    pool heaters, Jacuzzis, etc.):

device	hours of use per week
_____	_____
_____	_____
_____	_____

Finding the Price Tag

These estimates of your family's demand for various appliances within its five energy systems will become more useful after that demand is translated into dollar terms.

One way to make this translation is to look back on a few old utility bills. These bills answer two basic questions: they reveal how much electricity or gas you have used during the prior month, and they tell how much that electricity or gas costs per unit. If you keep bills for several years back, you might compare a very old one with a recent one, to see how much the rates have gone up in the area. Americans pay very diverse amounts of money for the same basic commodity. Electric rates fluctuate from three cents per kilowatt hour (the unit of measure) to triple that amount. It depends on which electric fiefdom you happen to live in.

The real cost of electricity or gas is often disguised in the bill behind various surcharges, taxes, and assessments—as illustrated in the typical utility bill, Plate 1. The best way to compute the actual rate is to take the total amount that you owe and divide by the number of kilowatt hours that you have used. (We will get to the meaning of a kilowatt hour farther along in this section.) In the example bill, the total amount is $40.31, which, divided by the 900 kilowatt hours used, gives a result of $.0448, or about 4½ cents per kilowatt hour.

For the cost calculations made throughout this book, we have chosen an electric rate of 5 cents a kilowatt hour. If you aren't paying that much now, the chances are that you soon will be. You may already be paying much more. In fact, New York City residents already shell out nearly 10 cents a kilowatt hour for electricity in the summer months. Five cents is close enough to what most people pay to make a good base figure for the energy budget. If you pay a different rate, you can adjust our numbers to fit your situation.

Electricity is sold in kilowatt hours; natural gas is sold in cubic feet. A gas bill is read the same way as an electric bill. Divide the total amount you owe by the number of units used (a unit is 1,000 cubic feet) and you arrive at the price you pay per unit. Types of petroleum derivatives, such as liquid propane or fuel oil, are metered at the delivery truck; your monthly statement should show the amount delivered to your tank, and the price. By keeping track of these propane or oil statements, you can know exactly how much you spend for cooking, or for heating, or for cooking and heating combined.

By adding up all your electric bills and gas bills and oil bills for the past year, you can compute your annual home energy *consumption*. If you live in an all-electric home, the total will be in the thousands of kilowatt hours (KWHs); if you use gas for cooking and heating, the yearly requirement for that fuel will be in the thousands of cubic feet. The *price* you paid for a year's energy is also an important thing to know. With the price tag in hand, you can translate the 15-percent reduction goal into a dollar amount, which should make energy conservation a little more inspiring. If, for example, you paid out $1,200 last year for electricity and fuel oil, then the 15-percent reduction will save you $180; and if you can manage to achieve a 50-percent reduction in usage, you get a dividend of $600 a year.

Marking Your Progress

If last year's monthly utility bills, taken together, can help you put a dollar sign on your energy plan, taken separately they can help you mark your progress through the plan. You can't wait all year to learn whether your conservation efforts are succeeding or failing. The old bills offer a point of comparison. A few utilities, like Seattle City Light and Power, recognize the importance of the comparison by printing last year's amounts on every utility bill currently received.

Since home energy requirements are different during each season of the year, you can't very well compare a September electric bill with a December electric bill. But you can compare this September's bill with the one you got in the same month last year. If you spent $55 for electricity last September, and only $49.50 this September, then you have achieved a 10-percent reduction. You can make such a comparison for each succeeding month. But there is one thing to watch out for: If electric rates have gone up during the last year—which they have a habit of doing—your conservation successes will be hidden in the higher rates. When rates increase by 10 percent, you have to reduce consumption by 10 percent just to keep your bills the same as the bills you got last year. You will still have saved the 10 percent, but you won't be able to tell that by looking at the bottom line on the bills. That is why it is better to measure your progress in kilowatt hours (KWH), or in units of gas, instead of in dollars and cents. If you burned 1,000 KWH last September, and 750 KWH this September, then you have reduced your electric appetite by 25 percent, and at 5 cents per KWH, you have saved $12.50 for that month.

There is one final advantage in reviewing old utility bills. If you live in an all-electric home, the old bills can help you separate out your costs for heating or cooling, and for all other appliances combined. Choose a month when you spend nothing on heating or cooling—say, October or May—and note the amount of money you paid on that month's utility bill. That is what it costs you to run the basic household, including lights, televisions, water heater, dishwasher, refrigerator, washer, dryer, freezer and stove. Then look at a typical bill for a winter month. The additional amount of that bill over the other one should provide an estimate of your heating costs. If the October bill was $40 and the January bill is $160, you can figure that about $120 went into heating during January. A similar calculation can be made for summer air-conditioning costs. A look at a few electric bills, therefore, can give a fairly accurate home energy profile: what you pay for heating (December, January, or February bills, less the amount for October) and what you pay for cooling (July and August bills, less the amount for May).

Write down some of the things you have learned from the old utility bills. The results will be helpful later.

(1) Last year's total energy bill (electric, gas, oil) $_____
(2) Winter heating costs $_____
(3) Summer air-conditioning costs $_____
(4) Basic household energy costs, for everything besides heating and cooling (October or May bill) $_____

Reading the Appliance Labels

Dusting off a few old utility bills is a good way to begin an energy program, but more information is needed. In a food diet, you can't lose weight until you know the fat-producing potential of everything you eat. In any energy diet, you must know the *debt*-producing potential of all the appliances you use.

There are two ways to attach a price tag to the various energy functions in an average household. One way is to read the electric meter. If you have such a meter, a daily peek at its dials will tell you not only how many kilowatt hours you have used, but which appliances have used the most. (We will explain how to read the meter later in this section.) Meter reading gives a complete energy profile of a house, a profile that can be invaluable to a successful energy-conservation effort.

If you don't have a meter (say, you live in a large apartment complex, and the meters are all tucked away somewhere in the basement section of the building), or don't want to read it, then you can look at the energy budget from another angle: at the appliances themselves. You probably have plenty of them. According to the reckoning of one utility company, there are thirty-two separate appliances in the kitchen, twenty in the dining area, twenty-nine in the living area, and twenty-one in the bedrooms, plus whatever you've got in the playroom or out in the garage. By learning how they affect the utility bill, you will know when to avoid using certain appliances and when to use others more frequently.

While small appliances carry labels that tell you how much electricity they use, most of the bigger ones —which really run up utility bills—do not have such labels; that is one of the reasons that people find it difficult to budget their energy. If a continuously running appliance does carry a wattage label, then figuring out the energy cost is simple. It takes 1,000 watts to make one kilowatt, so a 1,000-watt appliance, running for an hour, uses one kilowatt (KWH) of power.

A 1,000-watt portable hair dryer, therefore, costs 5 cents an hour to operate, at our estimated 5-cent-a-KWH rate. Ten 100-watt light bulbs, all burning for one hour, also consume one kilowatt. To convert watts into money, look at any appliance with a wattage label, find the amount you pay for electricity, and plug the numbers into the following formula:

$$\frac{\text{wattage}}{1{,}000} \times \text{electric rate} = \text{cost of operation per hour}$$

If you have an old 350-watt color television set, and you live in an area where electricity costs five cents per KWH, then the computation would go like this:

$$\frac{350 \text{ watts}}{1{,}000} \times 5 \text{ cents per KWH} = 1.75 \text{ cents per hour}$$

The problem with figuring the cost of operation for most large appliances is that they don't run continuously. Central air conditioners, central heaters, water heaters, refrigerators, freezers, and electric ovens all operate on thermostats that turn them on and off as desired temperatures are reached. The best we can do is to help you make your own calculations by filling in whatever information you have at your disposal. General estimates are meaningless. Since you have already computed the number of times you use large appliances each week, you are now one step ahead of the game. You can make your own calculations in the small log provided later in this chapter. With the information that follows, now, you should have a pretty solid idea of how much each of your five basic systems costs you to operate.

Calculating the Costs of Your Four Energy Systems

(1) The Hot-Water System

(A) *Water Heaters.* Many utility companies estimate that water heaters use about 4,500 KWH per year, or $225 at our 5-cent rate. Gas water heaters only cost between one-third and one-fourth that amount to operate—at least as long as prices are held down. The annual budget for heating water varies from place to place, however; and the other trouble with the estimate is that it is *annual*. And people won't want to cut down on water consumption until they know how much each bath, shower, or load of dishes is actually costing them.

You already know how many gallons of hot water you require each week. (Actually, you have computed only the amount of heated water that you need, since hot and cold are mixed together in showers, baths, and washing machines. But with the considerable heat losses in water tanks and connecting pipes, it probably takes nearly 100 gallons of purely hot water to make 100 gallons of the heated mixture you use from the faucet; so we'll assume that the two amounts are equal.) To make a cost estimate, you need to know only two other things: one is the temperature of the water as it enters your house, and the other is how high your water-heater thermostat is set. If you have an electric water heater, and are paying 5 cents a KWH, it will cost you 84 cents to raise the temperature of 100 gallons of water from 70° to 140°. So electrically heated water is very expensive. In areas of the north, water enters the house much colder than 70°, and many water heaters are set much higher than 140°. The greater the temperature differential, the more money you must pay. To raise 100 gallons of water by 100°, for instance, costs $1.20. (To determine just how cold the water entering your home is, allow the cold-water tap to run for two minutes, then fill up a pan with tap water. Put a thermometer in the pan and it will tell you exactly how cold the water is.)

Table 1: Electric Water Heater: Energy Cost

Temperature rise (cold to hot)	Cost to heat 50 gallons	Cost to heat 100 gallons	Cost to heat 150 gallons
70°	$.42	$ .84	$1.26
80°	.48	.96	1.44
90°	.54	1.08	1.62
100°	.60	1.19	1.79
110°	.66	1.32	1.98

Electrically heated water, as you can see, is not a commodity to be trifled with. It is one of the easiest places to save money, especially if you use a lot of it.

Heating water with natural gas is much cheaper. If you have a gas heater, you can figure your cost from this chart:

Table 2: Gas Water Heater: Energy Cost

Temperature rise (cold to hot)	Cost to heat 50 gallons	Cost to heat 100 gallons	Cost to heat 150 gallons
70°	$.09	$.18	$.27
80°	.10	.21	.33
90°	.12	.23	.35
100°	.13	.26	.39
110°	.14	.28	.43

(B) The Washing Machine. Most of the money goes to producing the hot water. It costs only about half a kilowatt, or 2½ cents to run the washing machine motor for an hour. At those prices, it isn't worth making further calculations.

(C) The Dishwasher. Same as the washing machine. The costs of running the motor are negligible.

(D) The Clothes Dryer. This is one of the biggies! Most electric dryers use about 4.8 kilowatts per hour. You can compute your cost this way:

$$\frac{\text{KWH}}{\text{per hour}} \times \frac{\text{Hours}}{\text{per week}} \times \frac{\text{Your electric rate}}{} = \frac{\text{Your cost}}{\text{per week}}$$

The Energy Budget: Staking Out the Meter

(2) The Heating System

Portable electric heaters are rated in watts, and they operate almost continuously when they are used, so it's easy to derive a cost of operation. For example, if you pay 5 cents a KWH and use your 1,500-watt heater for one thousand hours a year, the cost is:

$$\frac{1{,}500 \text{ watts}}{1{,}000} \times \text{KWH cost} \atop (5 \text{ cents}) \times \text{hours} \atop (1{,}000) = \text{\$75 yearly cost}$$

For other kinds of heaters, you will have to rely on the electric meter or on fuel bills to derive the cost.

(3) The Cooling or Air-Conditioning System

The same calculation we just used for portable electric heaters can be used for room air conditioners that are not thermostatically controlled. To make cost calculations for larger, central air conditioners, you'll have to rely on the test we present in the no-cost section of the book (Section 3).

(4) Cooking and the Kitchen

Natural gas is a much cheaper way to cook than electricity. If you cook with gas, the oven and range are not places where you will save much energy money, whereas electrics may be worth some attention. The large electric burners use about 1.7 kilowatts per hour, the small ones only use 1.1 kilowatts. Most ovens use 2.6 KWH, broilers 3 KWH, and the self-cleaning ovens draw about 2.8 KWH. So it costs about 15 cents an hour to broil, a little less than that to bake. You can multiply these figures by the hours you use the ovens and ranges in order to get a cost for cooking.

Refrigerators vary a tremendous amount in the number of KWH they require, as we discuss in the section on how to buy appliances (Section 5). We can't tell you how much electricity yours will use, but we can give you the range of prices you might pay for various sizes and types (these calculations are printed in the *1977 Directory of Certified Refrigerators and Freezers,* published by the Association of Home Appliance Manufacturers):

Table 3: Refrigerator: Energy Cost

Size in cubic feet	Cost of energy per month in dollars
9.5–12.5	$1.80–$2.40
10.5–13.5	1.80–$2.40
11.5–14.5	1.70–$2.90
12.5–15.5	1.70–$2.90
13.5 and over	1.70–$2.90

Table 4: Combination Refrigerator/Freezer: Energy Cost

Size in cubic feet	Partial automatic defrost	Automatic defrost
10.5–13.5	$1.90–$4.40	$4.60–$5.30
15.5–18.5	2.00–$3.20	3.40–$6.80
18.5–21.5		4.00–$8.20
21.5–24.5		4.90–$7.20
27.5–30.5		7.40–$7.40

As you can see, refrigerators can be major energy users, especially the big ones with automatic defrost.

(5) The Energy System for Lighting

Multiply your light hours times the average wattage of the bulbs you use in order to get a weekly cost figure.

(5a) The Energy System for Entertainment (and Miscellaneous Items)

Televisions use different amounts of power, depending on whether they are black-and-white or color, and solid-state or tube. Black-and-white solid-state types use about 60 watts; black-and-white tubes use 180;

color solid-state use about 165; and color tubes use 325. To figure your weekly cost, plug in to our wattage formula:

$$\frac{\text{watts}}{1{,}000} \times \frac{\text{hours}}{\text{per week}} \times \frac{\text{cost per}}{\text{KWH}} = \frac{\text{weekly}}{\text{cost}}$$

If you have waded through the above, you will now have a fairly accurate energy profile of your house.

Fill it in here:

		Weekly cost	Annual cost
(1)	Hot-Water System	_____	_____
(2)	Heating System	_____	_____
(3)	Cooling System	_____	_____
(4a)	Energy System for Cooking and the Kitchen	_____	_____
(4b)	Energy System for Lighting	_____	_____
(4c)	Energy System for Entertainment	_____	_____

Small Appliances

In addition to the large appliances, an endless number of small gadgets are found in most American homes. Since World War II, we have picked up most of these conveniences—primarily through birthday, holiday, and wedding gifts—at a rate much greater than one new appliance a year: toasters, broilers, electric ovens and spits, weed eaters, electric organs, aquariums with motors, electric knives, can openers, garbage disposals, digital clocks, phone answerers, electric lazy Susans, bar refrigerators, vibrators, hamburger-patty makers, popcorn poppers, exercisers, battery chargers, yogurt makers, food processors, bun warmers, ice makers, extra televisions, patio lights, stereos, Jacuzzis, massaging chairs, blow dryers, hair stylers, electric toothbrushes, and the instant hot. Most of these small items carry labels that reveal their

Table 5: Electric Power Consumed by Small Appliances

Appliance	Watts	Usage	Annual KWH	Annual Cost @ 5¢/KWH
GROOMING				
Hair dryer	1,000	1 hr. per week	13	$.65
Shaver	18	1 time per day	.54	.03
Sun lamp	300	10 mins. per day, winter	6	.30
COMFORT				
Blanket*	160	nightly, 6 mos. per year	120	6.00
Clock	2	continuous	17	.85
Fan, floor, 20"	160	6 hrs. per day, summer	115	5.75
Fan, window, 20"	200	3 hrs. per day, summer	72	3.60
Heating pad*	60	6 hrs. per month	4.4	.22
Oral water jet	120	4 times per day	14.4	.72
Toothbrush	1.1	15 mins. per day	.01	.01
Vibrator	40	1 time per day	1.2	.07
ENTERTAINMENT				
Hi-fi, tube-type	230	3 hrs. per day	248	12.40
Hi-fi, solid-state	120	3 hrs. per day	130	6.48

*Appliance thermostatically controlled.

Appliance	Wattage	Usage	Cost
Radio, tube-type	125	6 hrs. per day	13.50
Radio, solid-state	25	6 hrs. per day	2.07
TV, B/W, tube-type	180	6 hrs. per day	19.45
TV, B/W, solid-state	60	6 hrs. per day	6.50
TV, color, tube-type	325	6 hrs. per day	35.10
TV, color, solid-state	165	6 hrs. per day	17.82

CLEANING

Appliance	Wattage	Usage	Cost
Disposal	1,100	3 times per day	.90
Floor polisher	325	1 hr. per week	.85
Iron*	1,100	2.5 hrs. per week	3.60
Trash compactor	400	30 min. per day	3.60
Vacuum cleaner	700	15 min. per day	3.15

FOOD PREPARATION

Appliance	Wattage	Usage	Cost
Blender	350	5 times per week	.23
Broiler	1,150	3 times per week	3.60
Can opener	200	2 times per day	.15
Carving knife	100	3 times per week	.50
Chafing dish*	1,150	1 time per month	.75
Coffee maker	650	1 time per day	3.90
Corn popper	400	2 times per week	.52
Crock-pot*	250	2 times per week, or 8 hrs.	5.20
Deep-fat fryer*	1,150	4 times per month	2.10

Appliance	Watts	Usage	Annual KWH	Annual @ 5¢/KW
Frying pan *	1,250	1 time per day	134	6.70
Griddle *	1,200	1 time per week	31	1.55
Ice-cream freezer	130	2 times per month	1.6	.08
Ice crusher	210	5 times per week	5.6	.27
Juicer	125	1 time per day	3.3	.17
Knife sharpener	40	1 time per week	.2	.01
Microwave oven	650	30 min. per day	117	5.85
Mixer, hand	120	5 times per week	1.6	.08
Pressure cooker *	1,400	5 times per week	182	9.10
Roaster *	1,200	2 times per month	14.4	.72
Rotisserie	1,400	4 times per month	134	6.72
Sandwich grill *	1,150	5 times per week	149	7.48
Tea kettle	1,500	2 times per day	180	9.00
Toaster	1,400	1 time per day	42	2.10
Waffle iron *	1,320	30 min. per week	17	.85
Wok *	1,000	2 times per month	12	.60

* Thermostatically controlled (on for only part of the time).

wattages, so you can tell by looking which ones use more than a piddling amount of electricity. Most of them don't use much, and trying to cut a utility bill by unplugging a few of these small appliances is like trying to lose weight by not eating the last few peanuts in the can.

The *value* of small appliances is that they can often be substituted for the biggies that consume much more power. It is very uneconomical to toast bread in a regular oven when you have a toaster available. Similarly, a small electric blanket can save you a lot of energy money if it enables you to turn down the thermostat on the central heater. Such strategies will become apparent, once you compare the wattage ratings of the small appliances with the cost of running your big appliances. The following chart gives the estimated wattage ratings of various smaller items.

Reading the Meter

Keeping an eye on the wattage labels is one way to create an energy budget, but reading the electric (or gas) meter is even better. The meter not only gives you a much more accurate budgeting method, but it also provides you with a daily kilowatt count, a yardstick to measure energy-saving progress, and a reminder that keeps you motivated.

You may not have given your meter a second look since you moved into your house or apartment. A utility company operative peeks at it every month, but you don't know what he knows until you receive the bill several days or weeks later.

It is possible to put some of the blame for America's ravenous energy appetite on the placement of electric meters. Yours is probably somewhere around in back of the house, or, if you live in an apartment, in the basement. Such inconvenient placement was not a full-blown plot on the part of the utility company —eager to sell a few extra kilowatt hours—but you no doubt consume more electricity as a result. If the meter were in the kitchen or living room, where you could watch the dial go berserk when the air conditioner kicked on, you might quickly decide to use less air conditioning. (There is a device you can buy that is mounted in the kitchen and tells exactly how much electricity you are using throughout the household at any given moment. It will be described in Section 4.) And if people had to put actual money into their electric meters, as they do in parking meters, the country could probably sell half its electric generators and export some oil back to the Arabs.

You can take some of the mystery out of utility bills with a little spirited detective work: a daily visit to the electric meter. The results should be edifying. Reading your meter can help you in a lot of ways that we will get to later, but the first thing it can do is to guard against overcharges that result when the utility company makes a mistake. Since customers rarely check, misreadings can frequently occur; and some utility companies don't even bother to read the meter every month, but only make estimates that are often far off the mark. Getting the jump on the meter reader can be as felicitous as discovering a bank error in your favor.

There are two basic types of electric-power meters. The newer, simpler type shows four of five visible numbers. One look at the meter and you know how many kilowatt hours your home has consumed since the meter was installed. A reading of 14,500, for instance, corresponds to 14,500 kilowatts. A second kind of meter is more commonly used, but it is also more difficult to interpret. An example of this type of meter is shown as Plate 2 in the illustration section. It has a row of five clock-like dials, and a large horizontal wheel in the center. For the sake of clarity, we will refer to the dials—from right to left—as A, B, C, D, and E. The horizontal wheel is F. (Perhaps *your* meter may have only *four* dials, as do, for example, those of many thousands of apartments in New York City; in that case, you are missing, in effect, Dial E.) Each

The Energy Budget: Staking Out the Meter 37

dial turns in the opposite direction from the next one. If dial A turns clockwise (to the right), dial B turns counterclockwise (to the left), and so on, across the row.

The meter records the amount of power consumed in kilowatt hours (KWHs). A load of 1,000 watts on the line for one hour will use one kilowatt hour (KWH), or one unit of power. A 1,000-watt hair dryer, or ten 100-watt light bulbs, on the line for one hour will consume one KWH of power.

The dial farthest to the right on your meter, Dial A, registers single KWH units. Every time the pointer moves from one number to another on this dial, it costs you from 4 cents to 10 cents, depending on the rate you pay for electricity in your area. The meter is something like a cross between a clock and the odometer (mileage indicator) in a car. On a day when you are running the air conditioner or the electric space heater, you can stand and watch the money circling around the meter.

The second dial from the right, Dial B, indicates tens of KWHs; Dial C indicates hundreds of KWHs, Dial D thousands, and Dial E tens of thousands (if you have a Dial E). Dial A must go through an entire revolution (10 KWH) before Dial B is advanced by one number. A complete revolution of Dial B records 100 KWH, and advances the pointer on Dial C by one number. The five dials, read from left to right, show the total number of KWH used. In the example shown in Plate 2 that total is 14,099 KWH.

Fig. 1

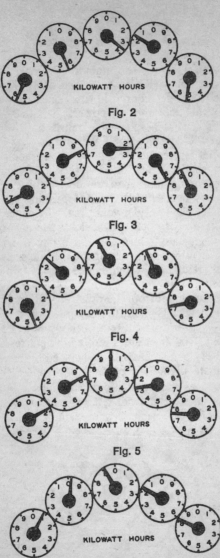

The Energy Budget: Staking Out the Meter

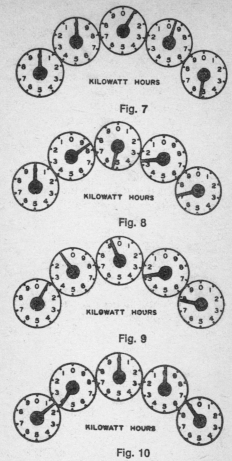

Fig. 7

Fig. 8

Fig. 9

Fig. 10

To test your skill at reading an electric-power meter, read the meters shown in Figs. 1-10 and record your answers on a piece of paper. Check your results against the answers. If all your readings are correct, you are ready to move on. If one or more reading is incorrect, reread the text, then read and record the ten meter readings again. This time you should get a perfect score.

Answers to electric power-meter readings above: Fig. 1—00010; Fig. 2—55315; Fig. 3—68259; Fig. 4—41907; Fig. 5—18027; Fig. 6—09918; Fig. 7—00095; Fig. 8—98527; Fig 9—00927; Fig. 10—13999.

The only difficulty in reading a meter arises when you are unsure which number a pointer indicates. You can resolve the confusion by looking at the dial to the immediate right of the disputed pointer. Let's say that the pointer on Dial C has landed near or on the number 4, but you don't know whether to read it as 4 or 3. Go back to Dial B. If the pointer on Dial B has passed zero ("12 o'clock"), then the disputed pointer on Dial C should be read as the higher number: 4. If the pointer on Dial B has not yet reached zero, then the pointer on Dial C should be read as the lower number, or 3.

A little practice should give you the idea. Read the total number of KWH shown on the ten meters. Answers are given, so that you may check your skill.

The gas meter can be read in the same manner as the electric meter. It usually has three dials, which register cubic feet of gas instead of KWH of electricity. The dial farthest to the right shows hundreds of cubic feet, the middle dial shows thousands, and the left-hand dial shows tens of thousands. Since your gas meter is probably not attached to as many appliances as the electric meter, you may need to read it only once a week, or even once a month. But if you do have meters, try to keep up with all of them.

The Benefits of Meter Reading

When meters are understood, they become tools for economy, as important as a thermostat or a water-saving toilet. You can read the electric meter every day, to find out exactly how much power you have bought over the last twenty-four hours. We suggest that you make an electric-power log sheet, such as Plate 3 illustrates, to fill in your daily readings. For continuity, it is best to read the meter at the same time each day. If you looked at 9 a.m. yesterday and the meter registered 14,199 KWH, and today at 9 a.m. it registers 14,249 KWH, then you have used 50 KWH of power in a twenty-four-hour period. At 5 cents per KWH, this means that your electric bill has increased by $2.50 during that twenty-four hours.

The Energy Budget: Staking Out the Meter   41

Meter reading opens up exciting new territory to the compulsive list maker, statistics lover, and calorie counter. It also has real practical value. For one thing, the meter can help you control your utility bills while you still have a chance to influence the bottom line; if you use too much this week, you can figure out a way to cut back next week. For another thing, meter reading provides a means to calculate whether your energy-saving efforts are working. You can check your progress every day by "weighing" your KWH consumption. And finally, a daily energy log can help show you which of your appliances, conveniences, and luxuries are costing you the most money, and where the most beneficial cutbacks can be made. The meter allows you to personalize your energy program.

Consider closely Plate 3, the log of a Florida family that has actually kept a record of daily meter readings for the past two and a half years. The entries for one month in that record, the period of October 27 through November 29, 1976, are shown.

This family uses a solar water heater, which reduces the overall electric bill by about 20 percent. Other than that, the log is fairly typical for an American home. The first thing to notice is that several of the days show about the same minimum amount of power consumed: between 14 and 17 KWH. Every household has such a minimum energy requirement: the amount of electricity or gas it takes to run all the *involuntary* systems like the refrigerator and freezer and water heater, to operate the television set, and to cook a couple of meals. On a day when this family doesn't use any major *voluntary* appliances, like the washer and dryer, and when they don't need heat or air conditioning, they consume about 16 KWH of power. At 5 cents per KWH, it costs them 80 cents a day to keep the basic life-support functions in operation.

After you have filled in your own log for a month, you can easily figure out your minimum energy requirement by looking at the days when you used the smallest amounts of power. Later, we will discuss ways in which this minimum daily usage can be reduced.

The second thing to look at in the sample log is the Remarks column. The family that made the log is careful to note any special events that require extra electricity. On November 21, the Sunday before Thanksgiving, the family cooked a twenty-two-pound turkey, which used 9 KWH of power and cost 45 cents to cook. On November 9, the speed grill was used to smoke chickens, which ran up another 15 cents on the meter. In the Remarks column, the energy manager also records the operation of appliances that are used only periodically, those like the washing machine, the dryer, the air conditioner, and the electric heater. You see that on October 28 the household used 28.5 KWH, quite a bit more than the minimum requirement of 16 KWH. Most of the extra energy is chargeable to the electric sprinkler pump, which uses 9 KWH over a three-hour period, at a cost of about 45 cents. On laundry days, the family uses 27 KWH, or 11 more KWH than it takes to run the basic household. That is 55 cents for laundry electricity. On November 25, the first cold day of the year, the family turned on the 1,350-watt space heater for four hours, which brought the day's total to 21 KWH. High and low temperatures are also recorded, which give calculations an added dimension, but it isn't imperative to record temperatures in your log.

The electric-energy log can help to ferret out certain energy "leaks" that you might not otherwise be aware of. This family was surprised to discover how much energy money one sprinkler pump required. Many people own similar "leaks": old freezers with bad gaskets, or oversized pumps—objects that they may immediately abandon when they discover the actual cost of operation. By eliminating the leaks, major energy savings can sometimes be achieved almost effortlessly.

After you have spent even a few days making the readings and noting them on your log, you will have a pretty good idea of how much various household activities cost your family. If you are careful to remember when large appliances are used, you will then know, for example, how much it costs to do the laundry, cook a turkey, bake some bread, run the heater. The daily amounts might seem small, but multiplied over a year they represent a lot of money. The information from your log is specific enough to enable you to make accurate judgments of where and how to cut down on electricity consumption.

Most people don't use every appliance every day, and as you can see from the sample log, the meter spins a lot more slowly on some days than on others. But during a week-long period, a family has probably gone through a complete energy cycle that includes the laundry, a big Sunday meal, etc. The *weekly* energy average can, therefore, be a focal point for your energy-saving plan. The log presented here shows that during a seven-day period between October 27 and November 3, the family used 160.5 KWH, for an average of 22.93 KWH per day. (This is a low average, because they have a solar water heater.) Every once in a while, you may have to revise your weekly average, especially if you use electric heating during the winter or air conditioning in the summer. During those months, your KWH consumption will obviously be much higher.

Once you calculate a weekly energy average, for whatever part of the year it happens to be, then you can begin the attack on your utility bill. Say, your average usage is 240 KWH for that week. During the next week—barring unusual weather changes—try for a 15 percent reduction, down to 204 KWH. Get your family together to discuss the possible ways to accomplish the reduction. Choose one of the appliances that you use the most, and then apply some of the no-cost suggestions listed in Section 3. You could even declare an "appliance fast," and use no major appliances for an entire twenty-four-hour period every week, as has been done with great success in one New York apartment building recently.

The point is, you can begin to experiment and see what works best for you. If you haven't gotten good results (according to the meter log) in one week of energy-saving, then try something else the next week. If family members aren't cooperating, try another form of motivation. At first the savings may be small, but as you go through the plan they should become much larger.

SECTION 3

Shedding the Load: How to Save Money without Spending Any

There are a lot of effective ways to save energy money without spending paycheck money. The best suggestions we have found fall into three categories.

(1) A few of the suggestions are "quick fixes," things you can do in a few minutes to achieve major savings. We put the quick fixes at the top of each system heading, so that if you are interested in instantaneous energy saving, you can thumb along to any section and pick them out at the beginning. And, after all, regardless of how much we may try to promote energy conservation through eternal, day-to-day vigilance, some of the most lucrative energy savers take only a minimal, once-and-for-all effort.

(2) Many of the suggestions are operational, and deal with how you use your appliances on a day-to-day basis. We have taken a lot of disjointed advice and tried to combine it into a veritable theory of operation for each important appliance. Learning a complete money-saving approach to the kitchen, in our opinion, will, for example, be much more profitable than learning new ideas on how to boil water.

(3) Some of the suggestions are structural, and involve a little maintenance. Thermostats can be reset, coils cleaned, and burners adjusted to get the maximum performance out of your various appliances for the minimum amount of fuel. Since you probably don't want to spend a lot of time poking around the back of the refrigerator or polishing burner reflectors on the stove, we have limited our list, however, to

those maintenance procedures that are demonstrably worth the effort.

The average American only employed one horse power in 1890 to help him or her with the chores. Now the average American employs the equivalent of 110 horses to do the same thing. The hitch in this arrangement is that we pay less than one-tenth the attention to our 110 "horses" than our predecessors did to their single horse.

Nobody is going to remember to change the air-conditioner filter today and check the refrigerator coil next week unless energy chores are made into a kind of ritual. The jobs can be displayed on a chart or calendar, so the delegated authority can remember what to do and when. To reduce grumbling, incentives can be offered, but the grumbling shouldn't be too intense, since all the chores won't take more than fifteen minutes a week.

Some of your home's or apartment's energy-using systems or appliances offer more potential savings than others, and if you have done the work in the preceding section, you will know which deserves the most attention. We can't tell you how much you will profit by each of the following suggestions, but we will provide an estimate of potential savings to give you a rough idea. (The estimates do not refer to the total savings for your total energy-saving plan, but only to what you can expect to get back from the no-cost suggestions in *this* section.)

Hot Water

Possible Annual Savings: $30 to $120

Some studies conclude that half the hot water produced in homes and apartments is wasted. We think you can eliminate a lot of the costly waste without spending any money to do so.

Electricity is an inefficient method of producing hot water, so if your system is electric you stand to benefit most from our suggestions. Gas water heaters are more economical to operate, but even they could stand some revision. Whichever way your water is heated, the tank itself is likely to be poorly insulated, which means that a lot of the heat is lost to the outside air, as we've said earlier. An additional amount of heat is dissipated in the uninsulated pipes that carry hot water to the far ends of the house.

Quick Fixes

Turn down the thermostat. One of the most lucrative adjustments in this book, in terms of payback for effort, can be accomplished by anybody with a screwdriver and a knowledge of where the hot-water tank is located. When a water heater arrives from the factory, its two thermostats are usually set between 150° and 160° Fahrenheit. The thermostats perform slightly different functions, but neither of them needs to be set at such a high temperature. The hotter the water, the more heat that is lost through the poorly insulated walls of the tank, and the more money you lose to the utility company. High thermostat settings can also damage the lining of the water heater.

It's easy to turn down the thermostats. They are protected by two metal plates located on the front of the water tank. Turn off the circuit breaker to eliminate any chance of electric shock, and remove the plates with a screwdriver. Push the insulation aside, and the thermostats should appear. By turning the pointers on the thermostats in a counterclockwise direction, you can reduce the water-temperature settings. You have to do it to both thermostats.

If you use a dishwasher, you should set the thermostats at 140°. That is the generally agreed-on temperature at which dishes get completely clean. If you don't have a dishwasher, a 120° setting should give you all the hot water your family will need. If you find that 120° doesn't give you enough hot water, you can readjust the thermostat slightly upward, say to 135° or 140°, and still save money over the original setting.

The economic benefits of the adjustment depend upon where you live. A study done at the Oak Ridge National Laboratory concluded that a thermostat setback from 150° to 130° will save 400 kilowatts, or $20 a year at out 5-cent rate. A setback from 160° to 120° will save a lot more than that. In either case, the adjustment will be worth the few minutes it takes to complete.

The Money-Saving Approach

Hot water is the circulatory fluid of the house. The demand for it in the average household is so great and so continuous that altering the *pattern* of that demand can save up to $100 a year, and can be even more profitable than modifying the hot-water system itself. There are two successful no-cost hot-water strategies.

(1) *Time your heater.* Reduce the amount of time the heater is in active operation. Some people save money by turning on the heater for only three or four hours a day, enough time to wash dishes, clean clothes, and take showers. Then they turn the heater completely off—until the next wash period. This time-limitation procedure works best with poorly insulated water tanks that require a lot of extra electricity just to maintain the water temperature over a twenty-four-hour period. Putting the water heater on a schedule will be especially effective when utilities offer night discount rates, a possibility we discuss in Section 11.

If you wonder whether this strategy could be worth the effort, we encourage you to try it on an experimental basis. After a day or two of limiting your hot-water demand—and reading the meter—you should be able to project monthly or yearly savings. If you have no meter, you will have to continue the experiment for a month and then analyze your utility bill to detect results. If the savings do not justify the inconvenience of putting everybody into the shower during a three-hour period, or of having to remember to switch the water heater on and off, we suggest that you insulate the water tank—a low-cost operation that is discussed in Section 4.

(2) *Reduce demand.* The second strategy is to reduce the amount of hot water used by your family.

The best way to do this is to restrict the number of times you wash (A) dishes, (B) clothes, and (C) take baths or showers. You can encourage people to be more judicious with the amount of clothes and dishes that they use during the week. Don't put out some sort of vague suggestion; figure out how many loads of wash you do each week, and try to reduce the number by a specific percentage, say by 40 percent. Encourage the children to wear old clothes after school, and the grown-ups to put on work clothes before tinkering in the garage or puttering in the garden. Make sure everybody has his *own* drinking cup, so you don't have to wash dozens of extra cups and glasses during the week.

The same scrutiny can be applied to baths and showers. You have already estimated the number of showers and baths your family normally takes in a week. Try to reduce that number by a fixed amount, and see if the results are visible on the electric-power log, or on next month's electric or gas bill. If results are visible, they will be a natural motivator to continue the effort. If you take a *short shower,* instead of a bath, every day, you save 3,600 gallons of water a year; four people would save 14,400 gallons. Heating 14,400 gallons of water from 40° to 110° requires 2,422 KWH of power, and at 5 cents per KWH, costs $121.10 per year—plus the cost for water and sewage treatment, which is no longer insignificant.

You can reduce the water requirements of a shower by taking it in a Spartan manner. Get wet, then turn off the water while you soap down; rinse off, and leave. Such austerity may never be popular, but became mandatory in the Western portion of the country during a recent drought. People who cut down on their shower time end up saving water and electricity; if people linger in the shower, they should be encouraged to take baths.

Another amount of hot water can be conserved if you don't shave using a continuously running faucet. In fact, the electric razor is actually an energy-saving device; the amount of electricity it uses is trivial compared to the amount required to heat the left-running water for lathering and razor shaving.

The economic results of hot water conservation will be most significant if you can reduce demand in all three areas—dishes, clothes, and baths—at once, and can combine the reduction.

(3) *Do full loads of dishes.* The dishwasher is one of those automatic appliances that actually saves money, compared to the reputedly ennobling manual process. A dishwasher usually requires 16 gallons of hot water. Unless you are very fastidious, you will use more than 16 gallons when you do the same amount of dishes by hand. The automatic dishwasher uses hotter water, but the human is likely to use so much *more* water that he loses the economic benefit of the lower temperature used in manual dishwashing. The automatic dishwasher, however, only remains economical if you do the full load, *and* only if you rinse the dishes in cold water in the sink before stacking them in the machine. By using the machine only when it is completely filled, you get the most from the 20 to 25 cents' worth of hot water used (the same goes for a clothes-washing machine).

Another way to save with the dishwasher is to let the dishes air-dry. Most machines have a heating element to speed up the drying process. If you do your dishes in the evening and let them dry overnight, speedy electric drying is unnecessary. Some of the newer models have a switch that will turn off the drying cycle; with older machines, you must shut off the dishwasher after the final rinse and before the drying cycle has started. Open the dishwasher door and let the air circulate around the dishes while they are drying. In the winter, the extra moisture released into the kitchen will help to add humidity to the house. And eliminating the hot-air cycle also has an added benefit in the summer: that extra heat will not have to be removed by your air conditioner.

If your machine has a Short Cycle control, use it; it reduces both the amount of hot water and the amount of power necessary to wash dishes. If you have a Rinse/Hold cycle, don't use it; it requires an additional 3 to 7 gallons of hot water per load.

(4) *Cooler-wash the clothes.* A washing machine should be operated on the coolest cycle that will produce acceptable results. A warm wash/cold rinse cycle will save money over a hot wash/warm rinse cycle. The Federal Energy Adminstration says washing machines could be made 23-percent more efficient by eliminating the warm rinse; and if you voluntarily eliminate yours, you should save that same 23 percent now. The Texas Power and Light Company reports that ". . . if the clothes were washed in 140 degree hot water with 110 degree warm water rinse, approximately 40 gallons of hot water is needed per load. That would require approximately 160 percent more energy than if a cold water rinse were used."

Hot water isn't recommended for modern synthetics and no-iron finishes, anyway. If colder settings don't get the clothes clean enough, you can try presoaking them in a cold-water solution. Do full loads, and try to wash on shorter cycles. Ten minutes should be sufficient for most types of clothes.

Two special energy-saving features can be employed on some clothes washers. One is the suds saver; it will allow you to recycle a tubful of hot, soapy water for another load of laundry—saving on soap money and hot-water money. The other is the mini-basket; it has a special cycle that requires less water.

(5) *Use the dryer less.* We include the dryer suggestions along with the hot-water system suggestions because the dryer is always used with the washer. The possible annual savings on dryer use are from $15 to $30 a year.

Most dryers consume about 5 KWHs of power to handle one load, and use about 25 cents' worth of electricity (at our average rate). The bulk of that electricity goes into the heating element, and not to the motor that turns the drum. The main point, as usual, is not to go off half-loaded.

General advice is complicated, of course, to a certain extent, by the types of fabrics you are drying. Some of the newer fabrics require lower temperatures and dry quicker when the machine is not packed full, so two short loads might take less total drying time and money than one very full load. You can experiment with your own dryer to find the minimum drying time required for various amounts of wash.

If you have more than one load of clothes to wash and dry at the same time, do the wash loads consecutively, so that the dryer will not cool down and then have to be reheated again. If you have slightly more than one load to dry, keep the small, lightweight items out of the machine. Those items can sometimes be dried later, on the heat which remains in the dryer after heavier pieces have been removed. The Fluff and Air-Dry cycles also save money, as they do not require power for heating the air.

If you do a lot of washing, major savings can be effected by partially drying heavy items and removing them from the machine to be solar-dried. The equipment required for this process can be purchased at very little expense at any hardware store; it is known as a "clothes line."

Try to calculate the exact point at which your clothes get completely dry. Leaving them in the dryer too long uses extra electricity and causes additional strain on the fabrics.

Removing the lint from the dryer filter before each drying session also increases the efficiency of the machine.

Maintenance

There are two ways you can maintain your hot-water system to make it perform more economically. Neither of these procedures offers anywhere near the return of the thermostat adjustment, but each can provide a small additional benefit if you are willing to put out the effort.

The first involves draining the water tank every few months. Sediment collects there and impairs the working efficiency of the tank. Open the plug at the tank base and drain a few gallons, until the water runs clear.

The second procedure is to watch out for small leaks in the various faucets around the house. A leak that produces a drop a second will add up to 300 gallons of water to your monthly consumption level. If it happens to be hot water, the 300 gallons could put an extra $4.20 on your electric bill.

Heating

Possible Annual Savings: $60 to $200

Quick Fixes

None.

The Money-Saving Approach

(1) *Lower the thermostat.* The most effective energy-saving device for whatever heating system you have is the thermostat you already own. Thermostat lowering may have gotten a bad reputation in your house after the government made 68° a matter of national policy. Such a standardized setting does not take into account the variations in humidity levels and local weather factors that make 68° acceptable in one place and much too cold in another. A lot of people made the patriotic gesture, tried 68°, shivered, and then sent the thermostat back up as high as it was before.

There are other rewards for resetting a thermostat, more tangible than the warm glow of patriotism. One study, done for the Department of Housing and Urban Development (HUD) estimates that a 6° drop in home temperature can reduce fuel costs by 15 to 20 percent. Whether this estimate would apply to your situation depends on a lot of factors that make such generalizations misleading. But you can use your own thermostat—along with the meter or the monthly fuel bill—to figure out what each extra degree of heat actually costs you. It may surprise you to discover that a tiny adjustment on the thermostat can represent many dollars over a winter season. If you read the meter every day, you may discover that slight alterations on the thermostat dial will begin to show up in the daily KWH count.

The easiest way to figure what a couple of extra degrees means to your fuel budget is to live for a week with a single thermostat setting—say, 72°. Then, try the next week at two degrees lower: 70°. If the average outside temperature remains about the same from the first week through the second, the results should be meaningful. Check your fuel costs after each week, either by reading the gas meter, looking at fuel-oil levels in your tank, or by monitoring the electric meter. If you have electric heating, figure the total number of KWHs consumed during the first week of the test; subtract from that amount the KWHs used the second week. The difference should represent the extra amount of electricity it took to heat the house at the higher temperature.

You can repeat the experiment for 68°, or for whatever setting you choose. By learning what each notch on the thermostat is worth in dollar terms, you can strike a balance between body comfort and financial reward. It is easier to adapt to a lower temperature if the change is made gradually, one degree at a time. Certain adjustments can be made to cause those lower temperatures to *feel* more comfortable, which we will describe later on. But the first way to feel more comfortable is to know how much money you are saving. It may warm your heart to realize that a lower thermostat setting is helping to finance a trip to the Caribbean.

(2) *The night setback.* Jiggling a thermostat back and forth doesn't help much in lowering heating costs; it's the long-term reduced setting that saves money. One of the easiest places to endure a lower setting is in bed. If the bed is warm, the rest of the house can stay cool until morning. A few blankets, a small bedroom heater, or even an electric blanket are much more economical to use than the big central heater or furnace. One study, cited by the Energy

Conservation Research Corporation, calculates the following benefits of various night setbacks:

Table 6: Fuel Savings by Various Night Setbacks of the Thermostat

Night setting	Percentage of fuel saved, 4 hrs.	Percentage of fuel saved, 8 hrs.	Percentage of fuel saved, 12 hrs.
68°	–	–	–
65°	1.5	3.0	4.5
60°	4.0	7.5	11.5
55°	6.5	12.0	18.0

When you turn the big heater down, and rely on blankets, bed partners, or small appliances to keep you warm, you are making money overnight. Try a few setbacks, and see how they affect your fuel bill. The only hitch in the setback plan is that somebody has to trek through the tundra to turn up the thermostats in the morning. (We will describe ways to avoid this trek in the following section.) The job can be passed around the family as a character builder.

You will also benefit by turning down the thermostat every time you leave the house for more than an hour. Some utility companies suggest that you set it at 55° when nobody is at home.

(3) *Zoned heating.* Some new houses deliver their heat in small areas, or "zones," so that the entire structure doesn't need to be enveloped in cozy extravagance. You can "zone" your own house, to a certain extent, by closing off the rooms that aren't being used. Make sure that windows and doors are shut in those rooms, and keep the heater registers completely closed. For rooms that are never occupied, you can even tape a piece of insulation over the registers. You can also encourage people to congregate in fewer rooms. The more bodies there are in a room, the more body heat can augment furnace heat.

The ultimate in low-cost zoned heating is the survival room, which limits the heated area to one section of the house during the coldest days of winter. The survival room is described in a later section.

(4) *Shut the vents.* Two energy studies, one done in Texas, the other in Colorado, both conclude that major air leakage into a house occurs around the kitchen exhaust fan and the bathroom exhaust vents, and also from the chimney of the fireplace. Dr. Jay McGrew, in Colorado, has even identified five major holes in the house that are responsible for greater increases in fuel bills, he says, than cracks around windows and doors. The five holes are: the furnace flue, the water-heater entry, the range vent, the bathroom-fan vent, and the clothes-dryer vent. You may save more money by closing off kitchen and bathroom exhausts and using the fans as little as possible than you will by investing in weather-stripping. The fireplace flue, of course, should be closed when you aren't using the fireplace.

(5) *Close the drapes.* Drapes make very good insulation if they are tightly fitted to the window frames. A test done by the Illinois Institute of Technology concluded that a medium-colored drape with white plastic backing would reduce heat loss by 6 to 7 percent in cold weather. The thicker the drapes and the tighter they fit around the window, the better they work as insulation. The dead air space created between the drapes and the window acts as a thermal barrier, to keep cold air from coming in and warm air from reaching the window and cooling off. Drapes have an advantage over permanent insulation because they can be moved. By opening the drapes on the sunny side of the house during the daytime, you will get extra heat in the house from that free furnace in the sky.

(6) *Bring on the fans.* Using fans in the wintertime may seem absurd, but fans can circulate the air around a space heater or small electric heater, spreading the heat to a larger area and creating more uniform temperatures. Experiment with a couple of fans in various locations and see if they help.

(7) *Humidify.* A dry indoor climate causes more water to evaporate from the skin, and the evaporation cools exposed necks, arms, and legs, and makes people feel chilly. When people feel chilly, they turn up the thermostat—which can introduce *more* dry air into the house, and escalate the whole process. One way to avoid this cycle is by using a humidifier, which helps people tolerate lower thermostat settings. A pan of water on top of a radiator or heater makes a good no-cost humidifier. So do moisture-producing chores, like mopping floors. During the winter, try to increase the moisture level in your house as much as possible by leaving the bathroom door open after you have taken a shower, by opening the dishwasher door and letting the dishes air-dry, or by cooking meals that require boiling water. Houseplants are excellent moisture producers, so if you are planning to adopt a few new ones, do it in the winter.

Some families even delay draining the bathtub until all the heat and moisture has been released into the house. If you think that's going a little far, an article by Roger Field in the October 1977 issue of *Science Digest* explains the benefits: "According to an expert at the International Solarthemics Corporation, you pay for approximately 10,000 BTUs to heat that water to bathing temperature. And pulling the plug immediately after bathing, in effect, flushes part of that investment down the drain." Field suggests letting the water stand for several hours before draining it.

(8) *Make heat.* Your baking, roasting, and extended cooking in the wintertime will also produce a lot of extra heat for the house. Leave the oven door open after you have turned off the oven and removed the cooked food, so that the heat can permeate the room. And if you really get inspired, you can pull up the vent on the clothes dryer, put a stocking over the end of the tub to catch lint, plug up the hole where the vent left the house, and let the dryer exhaust warm up your laundry room or bathroom. (DON'T do this with a gas dryer, as dangerous gases can leak into the house!)

(9) *Activate the people.* Sedentary bodies require more heat. When a body is working, it generates some of its own BTUs. When it is slumped in front of a TV, it depends more on the heat from the furnace. Television watching, for this reason, can increase the heating bill. If your family can become more active in the early evening hours, you may be able to turn down the thermostat. If they are willing to wear sweaters, you can turn it down even further. Or, consider the process in reverse: turn down the thermostat first, and people are certain to get more active after that.

(10) *The oasis concept.* The English have had centuries of practice living in chilly, drafty houses, where heat is found only in a tiny oasis around the beds, the fireplaces, or the kitchen stove. American homes once had such incomplete heating systems, or non-systems, in which people scurried quickly between the hot spots, rushing from their warm beds to put on their clothes in front of the pot-bellied stove or fireplace. In recent years, we have become accustomed to more than heating, we demand total climate control, in which a shirtsleeve level of comfort is maintained in all parts of the house at all times. This demand for climate control is a major cause of high fuel bills, especially in houses that are poorly insulated. When there are discrepancies in comfort levels between one part of the house and another, the thermostat is turned up high to bring the coldest part of the house up to an acceptable level of comfort; the rest of the house is hotter, then, than it needs to be.

Unless you modify your heating system with one of the energy-saving heaters described in a later section, totally heated environments will become economically unsupportable in the future. But before you consider a costly new heating system, you might want to consider a new approach to heating, away from climate control and back to the oasis concept. Such a change might require you to wear more sweaters, to use more blankets at night, and to limit the number of rooms you inhabit in the wintertime, but the economic rewards could more than compensate for the inconveniences.

Maintenance

Check the furnace. If you have an oil furnace, the seasonal checkup not only ensures against breakdowns, it also cuts fuel bills. Experiments have shown that a 1/15-inch layer of soot in the furnace can reduce efficiency by as much as 50 percent. There are too many kinds of furnaces for us to detail here; and the procedure of tuning them up can be dangerous for anybody who doesn't know what he is doing. The thing to do is to make sure a technician gets a look at your furnace before every heating season. A visit might cost a little money, but we put the suggestion in the no-cost section because a furnace checkup is part of regular home maintenance.

There are some things you can do to the heating system yourself. You can check the duct insulation, if any, for cracks or leaks. If you have radiators, keep the steam traps working properly, make sure nothing is blocking the registers, and avoid painting the metal surfaces. Paint makes a good insulator—and in this case you don't want one. Try to avoid the decorative boxes that cover radiators, thereby trapping the heat instead of releasing it to the room. If one of the radiators is located against a cold outside wall, put a piece of aluminum foil behind it. The foil will reflect heat back into the house.

Cooling or Air Conditioning

Possible Annual Savings: $120 to $500

Quick Fixes

Turn off the sump heater. This action won't give the best return of all our suggestions, but it will save a quick $20 a year. Your central-air-conditioning equipment probably contains a sump or crankcase heater, its purpose being to evaporate the liquid that might otherwise threaten the internal workings. As long as the air conditioner is connected to the electrical system of the house, the sump heater continues to operate. So, through the fall and winter months—long after the air conditioner has been turned off and forgotten—the sump heater keeps on using electricity. It draws between 50 and 75 watts of power. Assuming that a 75-watt sump heater is kept in operation through the seven months the air conditioner is *not* in active use, the total electric cost, at 5 cents per KWH, would amount to $19.50. If prices doubled, the sump heater would cost you $39 a year. The way to deactivate the sump heater is to switch off the air-conditioner circuit breaker at the end of the summer season. BUT: *You must remember to reset the circuit breaker at least twenty-four hours before you turn on the air conditioner for the next season. The extra twenty-four hours will allow heat to build up inside the mechanism and evaporate any liquid that might have collected there during the winter months. Without that "warm-up" time, you might damage the compressor of the air conditioner.*

The Money-Saving Approach

(1) *Increase the air-conditioner thermostat setting.* Many people first encountered air conditioners many years ago in movie theaters, where temperatures were kept at the goose-bump level to show off the new technology. Fifty years later, some people continue the tradition by setting home thermostats in the polar regions. It may therefore be easier, with no-cost methods, to cut air-conditioning bills than to cut heating bills. People generally prefer to be too warm than too cold, and, whereas pleasant summer indoor temperatures may be a matter of comfort, acceptable winter temperatures indoors are a matter of health. The ailment, remember, is called a cold (even though you *can* get one in warm weather).

A Southern family may discover that a 2° increase in their air conditioner's thermostat setting results in a one-third reduction in the monthly air-conditioning bill. Small upward nudges on the summer thermostat are just as lucrative as small downward pressures on the winter setting.

It's easy to discover how much you may benefit by altering your setting a little. If you have a room air conditioner that lists the number of watts on the label, you can tell, just by looking, how much it costs to run the machine each hour. Just remember that 1,000 watts equals one KWH—a nickel's worth of electricity at our estimated cost. If the label doesn't show watts, but does show volts and amps, multiply those two figures to find the watts: volts × amps = watts. If you can't find a label, or if you have a central air conditioner that doesn't list watts or amps, then do a meter test: Turn off all the electric appliances in the house and read the meter; turn on the air conditioner and run it for an hour; again read the meter, to determine how many kilowatts have been expended during the hour-long test. That's how much electricity the air conditioner uses in an hour of continuous operation.

Air conditioners, however, don't operate continuously, at least not the ones that have thermostats. They turn on and off, as temperatures are reached and lost again. There are complicated methods to time your air conditioner at various thermostat settings, but the easiest thing to do is to try one setting for an entire week, another setting for the following week; and then see how the meter readings recorded on your electric-power log were affected. Or, if you are careful not to use any other appliances that would throw off the calculations, you can do a daily meter experiment: Set the thermostat at 76° today, 78° tomorrow, and 80° the following day; and compare the KWHs used each day. Remember that such tests are not valid unless outside temperatures remain fairly constant. If you have no electric meter, the best thing to do is to live for a month with the thermostat at 79°, the following month at 80°; and then compare the utility bills.

And here are other ways to take a load off your air conditioner:

(2) *Use drapes.* They are effective in the winter-time, and even more effective in the summer. The drapes that reduce heat loss by 6 to 7 percent in the

winter can reduce heat gains in the summer by 33 percent, according to one university study. Drapes should be kept closed during the day, especially on the sunny side of the house. They keep sunlight from adding heat in places you are trying to cool. By opening the drapes at night, or whenever the outside temperature is *lower* than the room temperature, you can allow heat to escape from the house. Managers of a New York apartment building who embarked on an energy-saving campaign found that closing the drapes was one of the best ways to cut their air-conditioning bills.

If you have storm windows and also rely on air conditioning, leave the windows closed during the summer. Their insulative properties, which prevent the escape of warm air in wintertime, do the same for cold air in summer.

(3) *Close vents.* There is no justice, when you are creating an artificial indoor climate. During the summer, cold air leaves the house through the same openings where cold air entered during the winter. You can reduce these losses by plugging the same holes and closing the same vents that you sealed off during the winter. One thing is different: Whereas it helped to open the bathroom door after taking a warm bath or shower in the winter, it helps to keep the door *closed* during the summer, especially in humid climates. Moisture that adds to an already-high humidity increases summer discomfort and causes air-conditioning costs to rise.

(4) *Bring on the fans.* Fans are a very economical substitute for air conditioners. Large paddle fans use only the amount of electricity that a 100-watt light bulb would use. Room air conditioners have fan settings that require only a fraction of the electricity it takes to run the cooling unit. If the humidity outside isn't too high, you can turn on the fans at night, open the window, and forget the air conditioner. The benefits of this night shutoff will parallel the benefits of a night heating setback in winter. If you haven't tried a fan for a few years, you may have forgotten the comfort they do provide.

Portable fans can also be effective *in conjunction with* air conditioners. Fans can direct the flow of the cooled air and they can also cause people to accept higher thermostat settings. You might experiment with a few well-placed fans and see if they help you raise the indoor temperature with a minimum of discomfort.

Attic fans were once ignored, but they are now considered prime energy savers. Attic vents release the hot air that otherwise accumulates in the attic, and they reduce air-conditioning loads by a significant amount. We discuss how to install vents and attic fans in the next section, but if you already have such a vent or fan, you should use it.

(5) *Lay off the heat.* This advice may be trivial or important, depending on what kind of house or apartment you have and how much heat your daily activities generate. If you use a lot of heat-producing appliances, such as ovens and clothes dryers, try to schedule them during the coolest part of the day in order to minimize the extra burden they place on the air conditioner.

(6) *The dehumidifier principle.* Most houses are cooled like meat freezers. The inside temperature remains constant, regardless of what is happening outside. Weather becomes irrelevant, the changes in humidity outside during the course of the day are rendered unimportant—and the machine blasts away from morning until night. The meat-freezer approach has its advantages, but who can continue to support a 1,500-square-foot meat freezer at today's prices? A reconciliation with nature may be in order. We don't suggest the abandonment of air conditioners; most houses are not built to provide summer comfort without them. But by reacquainting yourself with the outside world, you can change the way you use the air conditioner—and at considerable personal profit.

We tested the "dehumidifier principle" on a typical Southern house, built with heavy cinder blocks and covered with a tile roof. The house is well insulated—important to this new approach to air conditioning. Since the house did not respond *rapidly* to temperature changes outside, we decided we could use those outside conditions to great advantage. The idea was to employ the air conditioner not as a constant source of cool air, but as a flushing device to remove humidity at specific points during the day. These points were determined by outside climatic conditions. During the nighttime and early-morning hours, the outside temperature stayed around 70°, and there was no requirement for air conditioning. (Some people might still want to use their air conditioner to circulate the air at night, but in the test house, we used fans to keep the air in motion.) The mass of the house, meanwhile, had gradually cooled off during the night—which added to our indoor comfort during the morning hours. The air-conditioner thermostat was set for 82°, and by the middle of the day, when outdoor temperatures approached 90°, the machine kicked on. We limited the amount of work it had to do by keeping the house tightly closed during this period, and the drapes were drawn. By closing up the house, we reduced the amount of heat and humidity that got in. The shaded, insulated interior felt very comfortable at 82°, as long as fans were used to circulate the air. Late-afternoon showers provided a natural air-conditioning effect. After the rains, the temperature dropped. We took advantage of the weather by opening the drapes when the outdoor temperature got below 80°. Since it was 82° inside, some of the extra indoor heat would escape to the outdoors through the now-uncovered windows. By nighttime, the temperature had dropped to the point where air conditioning was no longer necessary.

With the dehumidifier approach, the 4-ton air conditioner operated about three or three and a half hours a day. During that time, it consumed from 20 to 25 KWH of power, and removed between 20 and 25 quarts of water (humidity) from the air. It cost a little more than $1 a day, or about $35 a month, to use an air conditioner as a dehumidifier in the house we described. Other families in the area paid up to $150 a month for cooling their similar houses in the traditional way.

We don't know if the dehumidifier principle, as we have described it, will work in your house. We offer it as an example of the benefits that can be derived from using the outside temperature and weather conditions to best advantage. You have much more opportunity for such experimentation in the summer than in the winter. By paying attention to the weather patterns in your area, and by opening or closing your house at specific times during the day, you may find a way to make nature pay some of your utility bills.

Maintenance

The yearly checkup. Regular maintenance of a central air conditioner might save some energy money (although probably not as much as the maintenance done on a furnace). Loose belts and low coolant levels can cause a machine to draw more electricity. A service person can perform a test to see if your machine is using more power than it should. If it hasn't been tested in a while, you may discover that it is wasting money.

You can check the filters yourself; these can be removed from the return air ducts and dusted off each month. Shake them a few times to eliminate the accumulated dirt. If they are still clogged, and if you can't see through them, the filters need to be replaced.

Cooking* and the Kitchen
Possible Annual Savings: $20 to $130

Quick Fix

Douse the pilots. If you have a gas stove or range, you may be able to cut from 30 to 50 percent off the fuel bill without changing your cooking habits at all. The pilot lights have been shown to waste more than one-third of the fuel consumed by the stove. If you don't cook very often, you may be paying more for

* Consideration of the refrigerator/freezer part of the kitchen system follows.

your pilot-light fuel than for cooking. If your gas stove has two pilot lights, you may be wasting up to 9,600 cubic feet of gas, at a cost of over $20 a year. And think what the cost may be, when gas is de-regulated.

Some stoves have small adjustments that will eliminate the pilot lights. You may be able to turn off your oven pilots and activate them only when you are about to use the oven; if you turn off the stove-top burner pilots, however, then you will have to light the burners with a match. You will also have to be careful not to turn on a burner without lighting it, as gas could then escape into the kitchen and possibly ignite. These warnings given, some people are accepting the inconvenience of matches and the extra care required, in return for lower fuel bills. Some experts argue that pilot lights are not wasteful, since they add heat to the house during the winter; but these people fail to point out that during the summer the air conditioner has to work harder to remove that extra heat. And this gives us an economic standoff on the *secondary* thermal effects. But when cooking efficiency is considered alone, pilot lights are good things to get rid of.

Since gas stoves are potentially dangerous, you shouldn't tamper with any of the settings without getting help from a local appliance dealer or serviceman. What's more, your stove may not have adjustable pilots —in which case you might decide to convert to electronic ignition, a possibility discussed in the next Section 4.

The Money-Saving Approach.

John Fechter, of the National Bureau of Standards, wanted to investigate kitchen appliances, and ended up investigating cooks. He was pursuing the "hardware angle," or how you measure the energy-saving potential of various kinds of stoves and ovens. The idea was that stoves might be modified to conserve energy, and Fechter and others had invited some local cooks into a standard American kitchen, to put that idea to the test.

It was a normal kitchen setup, with the usual appliances; and all the cooks produced identical meals.

Fechter supposed that the various cooks would use the same amount of electricity or gas to make the meals, as long as they were using the same appliances. But he surprised even himself with the following conclusion: "When comparison was made on a meal-by-meal basis, the difference in energy consumption by the cooks can be more than 50 percent for the same meal, same range."

Fechter devised a new study: to discover why two chicken casseroles or two pot roasts, prepared by different people, in the same oven, could produce such disparate utility costs. He watched the various cooks move along the preparation triangle: from refrigerator to stove to sink. The cooks—housewives from the neighborhood—performed the kitchen ballet in a number of styles. One might open the refrigerator door twice, while another would open it ten times, to remove the same ingredients needed for the meal. One cook might leave the oven door ajar, testing the meat, while 450° heat enveloped her face like a sauna bath. Another might use the large burner for a tiny pan, or prepare sauces in two pans where one would do just as well. All these seemingly trivial factors added up to a notable difference in energy use. Cook A put a meal together for 3.4 KWH, while Cook B made the same meal for 5.2 KWH. The energy-conscious cook saved 10 cents on that meal alone. Over one year's cooking, the savings add up to between $20 and $30. That's not the big dividend you get from lowering a thermostat, but it does cut another little corner off the utility bill, and requires no attendant discomfort.

Cooking habits can't be changed overnight, and many cooks would gladly continue to pay the extra $20 to $30 for the privilege of doing things their way. But if you can begin to modify your kitchen behavior now, you will be better prepared for the day when fuel prices double and inefficient cooks pay a much *higher* penalty.

Most cooking tips offer trivial savings, when considered one at a time. As Fechter discovered, it is the combination of little actions that produces higher energy costs per meal. We list a few of the important tips in the hope that you can combine them together into a new pattern of cooking:

(1) *Don't preheat the oven.* A famous "hamburger study," done for the American Home Economics Association, and published in the June 1977 *Home Economics Research Journal,* measured how much fuel it took to cook a hamburger in various ways. The study concluded that preheating the oven is more costly than cooking from a cold start. If the recipe doesn't require preheating, then don't do it. You can, however, reduce cooking time by thawing frozen foods in the refrigerator (removing from freezer), or at room temperature, before you cook them.

(2) *Use the right burner.* The small burners on the electric stove require about 1,000 watts, while the large ones use 1,700 watts. When you put a small pan on a large burner, you are consuming nearly twice the energy that is necessary to do the job. Much of that extra energy is lost.

If you have been trying to decide which part of the stove offers the cheapest way to cook any given meal, the "hamburger study" offers the following advice: Cooking on top of the stove is cheapest, followed by broiling. Baking is the most expensive way to cook.

(3) *Don't peek.* Somebody figured out that a peek causes a 25° drop in oven temperature. Eight or ten peeks can be the equivalent of heating the oven all over again. If you have a meat thermometer, use it. If you must slice into the meat, do it when the pan is outside the oven and the door is closed.

(4) *Coast the roast.* Turn the oven off thirty minutes before the meat is done, and the heat you have already paid for should do the rest of the work.

(5) *Use the right pan.* If you use glass or ceramic baking pans in the oven, you can lower the recommended temperatures by 25°. Cover your stove-top pans whenever possible; a covered pan retains more heat and cooks faster. If you have a ceramic-top stove, it is even more essential that you use the proper pans. If they are not absolutely flat, much of the heat

produced by the stove top will be lost before it gets to the pan.

It takes about five minutes for a ceramic-top stove to warm up to full heat. You can compensate for this preheating period by turning the burner off five minutes before cooking is completed; the retained heat will finish the cooking.

(6) *Boil less water.* If you are boiling water for one cup of coffee, boil *one* cup of water. You get the coffee much faster and you save electricity or gas. Boil vegetables like potatoes with the smallest amount of water that will do the job; corn can be steamed with tiny amounts of water in a covered pan, instead of being boiled in a huge potful of water. Furthermore, a slow boil cooks just as well as a violent one, and at much less cost.

(7) *Use the thermos.* An insulated container like a thermos bottle will keep your coffee warm for free. Using an electric or gas burner for that purpose can get expensive.

(8) *The one-pot principle. How* you cook is not nearly as important to saving energy money as *what* you cook. Recipes are not yet expressed in energy terms, like the 1-KWH chicken, or the 4-KWH lasagna, but it may eventually come to that. Some countries have always considered fuel availability as part of the cooking process. China is one of them.

Since the Chinese populate an area of the world where fuel is scarce, they have developed a cuisine that uses little of it. Chopping meats and vegetables into small pieces makes them cook faster. Stir-frying food in a wok, which takes only a few minutes, produces a full meal on a tiny amount of energy. Steaming, which requires only small quantities of boiling water, is often substituted for baking or boiling; and, with several-tiered steamers, the Chinese can use the steam that rises through the tiers to cook several dishes at one time.

You may not want to eat Chinese meals all the time, but some of the Chinese principles can be applied to other styles of cooking. The one-pot meal is one of the best ways to economize on fuel. If you are cooking

Shedding the Load

outdoors, you try to put everything on the same fire; it is too much trouble to make two fires. Yet people routinely use three or even four "fires" in the kitchen, when they could get by with one. Meals like stews and chowders can be easily produced in a single pot. It is also possible to cook vegetables in the broiler, alongside the meat.

As long as you have paid for the heat, you might as well use it for more than one thing. When you heat the oven, cook several meals at once and freeze some for another time. If you are baking bread or cakes, bake several things and freeze what you don't need now.

(9) *The wedding-present principle.* You can also benefit by cooking with small appliances. The best way to decide which to use is to compare the wattage label of a small appliance with that of your larger one. Small electric broilers, for instance, draw about 1,500 watts of power, while the broiler in a standard electric oven uses about 3,000 watts, or more than twice as much. With gas stoves and ovens, the difference will not be as great; since it is cheaper to cook with gas, the advantages of small electric appliances is, therefore, limited.

It is highly unprofitable to toast bread in the oven instead of in a toaster. Toasters require about 5 cents an hour to operate, while it can cost three times that much to heat up the oven cavity. Many utility companies, in their "tips" booklets, suggest that electric frying pans are more efficient than the stove-top burner. The heat coils are imbedded inside the pan, so no heat is lost in the transfer between burner and pan. There is, however, some disagreement on this point. The "hamburger study" concluded that for *one* hamburger patty, the stove-top burner cooked cheaper than the electric skillet. But it may be more economical to use the electric skillet if its heating surfaces are fully occupied —by four patties, for instance.

Another inexpensive way to produce a meal is with a slow-cooker, or crock-pot. You can put food in the crock-pot in the morning, let it cook throughout the day, and serve it at night. Some crock-pots use a thermostat, others simply have very small heating elements; both kinds operate on very little power. In a test for this book, our monitored crock-pot cooked one gallon of stew for a nine-hour period, and consumed 1.2 KWH, or 6 cents' worth of electricity (at the 5-cent rate).

Whereas the crock-pot saves money by cooking very slowly, the microwave oven achieves the same results by cooking very fast. Your authors are not agreed as to the safety of the microwave oven; however, they do agree that microwave is one of the most economical ways to cook.

An additional advantage of small appliances is that they can be moved around. In the winter, of course, it makes sense to cook in the kitchen, where the heat given off by various appliances and the oven helps warm the house. But in the summer, kitchen heat produces costly thermal "collisions." These can be avoided by moving the crock-pot or the electric frying pan to an outside porch, or to some place away from the living area, where they won't fight the air conditioner. Even if you use them in the kitchen, however, the small appliances will add less heat to the house than will your oven or broiler.

Barbecuing is a popular summer alternative to heating up the kitchen, but the cost of charcoal and lighter fluid combine to produce one of the most expensive methods to cook food. There are cheaper ways to barbecue. With an electric grill that heats up lava rock, you can cook a round of steaks for about 4 cents; a barbecue grill that uses only newspapers for fuel will be described in the next section of this book.

(10) *Cold-dish meals.* Besides migrating with the cooking appliances during the warm season, you can also make more cold-dish meals. Such meals were a part of the summer routine in the days when air conditioning was not readily available. With no air conditioning in the house, the extra heat of a heavy cooked meal made people feel uncomfortable. With the air conditioner, we tended to come to discount those effects. But since the heat produced in the kitchen makes the air conditioner work harder, we now pay in the cost

of KWHs instead of discomfort for the privilege of eating hot meals in the summer. An energy-saving summer diet would emphasize cold soups, like gazpacho, and salads.

Maintenance

(1) *Keep reflectors clean.* This seems trivial, but there aren't many other ways to save money on stove maintenance. If burner reflectors are shiny, they will send more heat back up to the pan, where it is needed.

(2) *Clean the oven after cooking.* If you have a self-cleaning oven, do the cleaning right after you use the oven. The heat still left in the oven will give the cleaning element a head start.

Refrigerator and Freezer

Quick Fix

Unplugging the oldies. The meter log may have helped you discover your energy leakers, the big marginal appliances that might be costing you $100 a year to keep in operation. The second refrigerator or deep freezer is usually the principal energy leaker in a household. We mention it here because getting rid of it is the easiest and cheapest way to cut substantial chunks off your utility bill. A group of energy consultants who call themselves Living Systems analyzed the residential energy requirements of several houses in Davis, California. They found that refrigerators and freezers cost so much money to run that ". . . the question should be raised as to whether the benefits of bulk buying offset the costs of storage."

If you have a freezer, it may be worthwhile to recalculate the costs of electricity to run it against the savings you get from buying in quantity. If you have a second refrigerator—especially if it is only used to store marginal items like ice or extra beer—you might consider unplugging it. Such a refrigerator is usually relegated to the basement and kept for economy's sake—who would want to throw away a usable appliance?—but it turns out to be an economic burden. New refrigerators and freezers may cost up to $100 a year in electric bills, but if your old one has a bad gasket or an inefficient compressor, you may be paying more than that for its operation. If you don't believe this, try disconnecting the second refrigerator for a month, and see how it affects your meter log or utility bill.

The same procedure can be applied to large sprinkler pumps; to the second stove, especially if it runs on gas and has a continuously burning pilot light; to the superfluous air conditioners; and to unnecessary outside lights. Try to get rid of them, or trade them in for an energy-saving improvement.

The Money-Saving Approach

(1) *Stop the stare.* There isn't much else to do with a refrigerator except open it—which you may do too often already. But every family seems to have one or more members afflicted with the "refrigerator stare": that mesmerized pose, held through minutes of mental indecision while the door is open and the BTUs are escaping by the millions. The refrigerator stare is often brought on by wishful thinking—what you really want isn't there, but if you keep looking it will somehow appear. Such fantasies cost money.

The authors conducted an experiment in which they let a standard refrigerator run for a certain period of time without opening the door; then they repeated the experiment while opening the door every ten minutes. The closed refrigerator consumed 8.4 KWH, costing 42 cents for a full day's operation. The refrigerator that was frequently opened used 14.85 KWH, or almost double that amount. It is unlikely that your family's refrigerator will be opened as often as our tested one, but the experiment does prove that excessive traffic in and out of the refrigerator can have a momentous upward pull on the electric bill.

(2) *Defrost it.* Frost can really cut down on the efficiency of the refrigerator. When it gets more than a quarter-inch thick, it's time to defrost it.

(3) *Let it cool.* Give hot food enough time to cool down to room temperature before putting it in the refrigerator. You'll create no health hazard by letting small amounts of food cool off naturally.

(4) *Don't overload.* Don't cram things into the refrigerator. In a refrigerator where the freezer is a separate compartment, it is best to store perishable items in the back of the bottom shelf, because that is where it is coldest. In a refrigerator where the freezer is a part of the refrigerator compartment, the coldest place is on the top shelf of the refrigerator. That way you can use a less cold setting than if the perishable items are in the coldest part of the refrigerator.

In the freezer, it is better to pack things tightly together, and not leave too much air space between them.

(5) *Place it well.* Try to locate the refrigerator in a fairly cool part of the kitchen, away from direct contact with the stove and away from a western or southern outside wall.

(6) *Don't sweat it.* Many refrigerators have a heater in the door to reduce sweating during periods of high humidity. Door sweat may be less important, though, than the money you save by *not* using this door heater. If your machine has a switch to activate this heater, by turning it off you can save as much as 16 percent on electricity.

(7) *Turn it off.* When you go away for a few days, take the perishable items out of the refrigerator and raise the temperature setting; or better yet, turn the thing off. A large frost-free refrigerator, remember, can cost $8 a month to operate. Empty it, wipe it out with baking soda and warm water, and leave the door open.

Maintenance

(1) *Clean the coils.* The cooling coils, located on the back or under the refrigerator, should be brushed and dusted every two or three months. You should also make sure that the refrigerator is not jammed

against the wall so that air can't flow freely around the coils.

(2) *The dollar-bill test.* Another thing to check every now and then is the rubber gasket around the door. When it does not seal tightly, the outside air leaks into the refrigerator box, frosting up the coils and making the compressor work harder. A bad gasket can double or triple the cost of running the machine.

You can test your gasket by putting a dollar bill between it and the door at various points along the four sides of the door, then shutting the door. If the dollar bill can be easily pulled out, the gasket is defective. We will describe how to *replace* a gasket in the next section.

Lighting

Possible Annual Savings: $10 to $40

Quick Fixes

(1) *Cut the wattage.* When a bulb burns out, try to replace it with one that has less wattage. People often buy light bulbs that are more powerful than necessary. Every time you cut the wattage in half, you have cut 50 percent off the utility bill for that particular fixture.

(2) *Take one out.* If you have a fixture with two or three bulbs enclosed in a large globe, take out one of the bulbs. (Be sure to put a burned-out bulb in the socket hole, for safety.)

The Money-Saving Approach

Electric lights are symbols of the whole energy-saving program—visible, like litter is to ecology.

People have always been told to turn off lights, without having much idea of the potential return of doing so. But, unlike some other electric products, light bulbs tell you exactly how much power they consume. You don't have to read a meter or hunt

Shedding the Load

for a label. The wattage is stamped on the package and on the bulb. If a bulb is marked 100 watts, that means it will take ten bulbs, burning for one hour, to reach one KWH; and at our estimated rate this will cost a nickel. One 100-watt bulb burning for ten hours, of course, amounts to the same thing.

Chances are that four or five rooms are occupied at one time or another during an evening in your home. You have perhaps arranged a lighted path to the refrigerator—two or three bulbs along the way. There is maybe a light near the television; a reading light somewhere; or another lighted area in the living room; one in the game room, or one in the bathroom. Somebody may be in a bedroom reading or playing a musical instrument. An outside light is perhaps left on for the stragglers coming home. If you multiply the number of lights in operation by the total number of evening hours, it is not impossible to consume forty to fifty light hours per evening. You may not use that much; you may use more.

Room-to-room migration, too, can also get expensive—especially in a big house—because some people manage to go from one room to another leaving light bulbs burning like Sherman left fires in Georgia. If you are a migrator, it should be fairly easy to cut your light bill in half.

(1) *The automatic turnoff.* Get into the habit of never crossing a door jamb without flicking off a light. If there is no switch, then *pretend* to do it.

(2) *Turn out constantly burning, unnecessary lights.* Consider a home in which two 100-watt bulbs are burning unnecessarily for ten hours a day. They don't have to be in the same lighting fixture, and it is not unusual to have two lights operating in unoccupied areas of the house for ten hours each day. At 5 cents per KWH, the price for that excess comes to $36 a year, and if you can manage to keep them turned off, you have earned yourself a dinner out, or even a few movies.

(3) *Concentrate the light.* Some homes are lit up like border crossings. If you can reduce the general illumination and concentrate the light in smaller areas,

such as over a reading table or a kitchen table, then you save substantial amounts. And for daytime activities, of course, sunlight is still the best buy.

(4) *Use fluorescents.* A 40-watt fluorescent bulb provides more light than three 60-watt incandescent bulbs, and can save $10 per year. The only problem with fluorescents is that their useful life is drastically reduced by constantly turning them on and off. Put them in places where constant switching is not required.

(5) *How to buy bulbs.* Consider two factors: the lumen rating tells how much light the bulb gives off; the wattage rating is how much power it uses. Get the best lumen-for-watt deal that you can.

Buy low-wattage bulbs for night lights. Stay away from the 3-way bulb; in our experience, people always keep them on the highest setting, anyway. Long-life bulbs are not that good a buy, either; they last more hours, but they use more electricity per lumen than regular bulbs—what you win in longevity you lose in electricity.

(6) *Outside the house.* Turn off decorative lights when you can. People should be given awards for Christmas displays that DON'T use any lights. For safety lights, consider mercury vapor or some other energy-saving type. If you happen to have a gas lantern, removing it from service may save $25 a year. Converting it to electric will save $19 a year, according to the Federal Energy Administration.

Entertainment (and Miscellaneous Items)

Possible Annual Savings: $25 to $50

Quick Fixes

(1) *Check the swimming-pool pump.* If you have a swimming pool, you may want to investigate ways to reduce the time that the pump must operate. (Not enough people have swimming pools, however, for us

to devote space to these adjustments. But see "Saving Money on Water," too.

(2) *Eliminate the Instant-On.* Many of the old tube-type television sets were sold with an exciting feature called "Instant-On." When you turned the switch, the picture appeared almost instantaneously, whereas without this feature, the set would require from thirty to forty-five seconds to warm up. Instant-On is nothing more than extra electricity that is fed into the TV tubes twenty-four hours a day, 365 days a year. It costs a family that uses its TV only six hours a day about 486 KWH, or $24.30 a year, to enjoy the benefits of Instant-On. If you want to save $24.30, there may be a switch on the back of your set that will inactivate the system. It should be marked "Instant On–Off." If not, it is a simple matter for the TV repairman to cut the wire and eliminate the system that way. Or you can merely unplug the set every time you finish watching it.

Maybe you're not sure if you have Instant-On, or not. First of all, it only exists in sets with tubes. Look into the back of the set after it has been turned off a few minutes. If you see a glow in the small tubes and in the picture tube, you have Instant-On; you are paying extra money for a quick picture.

The Money-Saving Approach

Use a black-and-white TV instead of color. Black-and-white television sets use about half as much electricity as do color TVs. For the auxiliary set in the children's room or the game room, black-and-white could save you quite a bit of money. How much money, of course, depends upon how much television you watch. A black-and-white set with tubes uses 180 watts, or about 9 cents every ten hours; a black-and-white transistorized television set uses only 60 watts, or one-third the amount used by its tube counterpart. A color TV with tubes consumes about 325 watts. For purposes of comparison, the color TV with transistors is just about as economical as the black-and-white TV with tubes. A color TV with tubes costs twice as much to operate as the transistorized color model, and about six times as much as the transistorized black-and-white.

If your family watches TV for, let's say, six hours a day, the relative yearly electric costs would be:

B/W, Transistor	$6.50
B/W, Tube	19.45
Color, Transistor	17.82
Color, Tube	35.10

For people who use their television sets only intermittently, the energy costs are not very important. But for TV addicts who watch ten hours a day, or more, the yearly operating cost for one of the old tube-type color sets can exceed $100.

Special: Saving Money on Water

We have examined water only in the context of the cost of fuel to heat it, but there are good opportunities to save money by the conservation of the water itself. The U.S. Geological Survey recently reported that the average American family now consumes an incredible 172 gallons of water per day. The figure has been steadily creeping up in recent years. It might seem outlandishly high, but not when you consider the breakdown: a flush of the toilet takes from four to six gallons; every bath and shower, as we have seen, requires twenty to thirty gallons; the laundry takes thirty to forty; the dishwashing, fourteen; and when you add the car washes and lawn waterings—and the occasional leaky faucet—then a lot of water has gone down the drain.

In the past, water has neither been expensive nor in short supply. People haven't had to give it much thought. But the increased demand, coupled with higher costs for fuel and with construction costs for sewage and water-treatment plants, are beginning to change that. Every gallon that you take from the city line and then send down the drain creates a double cost; the cost of producing it, and the cost of treating and disposing of it. You either pay directly through high water charges—some areas like the Florida Keys are now selling water to the inhabitants for $2.75 a thousand gallons—or indirectly through high sewer- and water-bond issues. The high cost of water is making conservation a worthwhile object of consumer attention.

In some areas of the country, the *price* of water has not been as troublesome as the *supply*. Through extended droughts in the Western states, water has already gotten as scarce as fuel threatens to become. Inspired by voluntary or mandatory municipal conservation programs, people have been able to save remarkable amounts of water. They have learned to cut down on car washings and lawn sprinklings, and have adjusted to shorter showers, water-conscious dishwashing; and have even flushed the toilet only for solid wastes. The success of water-conservation crash programs suggests that water usage in the home is quite flexible, and that reducing consumption is not difficult.

Other areas of the country have not yet applied the Western experience to their own situations. Cities in Florida complain that lack of water may force them to curtail development, but they have not undertaken conservation programs to reduce their overall need for water. Municipalities that must purchase costly new sewage-treatment plants or water distribution facilities usually do not consider the possibility of cutting down consumption at the household level and thereby eliminating the necessity for new plants or facilities! Consumers have not examined their own residential water usage and the easy ways they might cut down.

One exception is the Washington Suburban Sanitary Commission, which provides water and sewage treatment to a large area on the outskirts of Washington, D.C. That commission did not want to wait for a drought, or a water-shortage crisis, before learning how consumption could be reduced. They started, in 1971, with an ambitious program of consumer education. They sent out fliers and handbooks; they asked their customers to suggest good ways that people might use less water; they conducted a local media campaign.

For those efforts, in the first stages of the program, the people of the area saved 1.7 million gallons of water.

After that, the commission distributed kits containing a dye, which helped consumers detect leaks in toilets. About 15 percent of the people who received kits did find and correct such leaks, which can waste large quantities of water. The commission also made up a guidebook listing manufacturers of control devices (which we discuss in the next section of this book), and even bought thousands of shower-flow controls, sold them at cost to householders, and instructed them on installation of the controls. As a result of these continuing efforts, per capita water consumption dropped from 101.6 to 97 gallons per day. Over the entire area, even that small change in per capita usage resulted in savings of millions of gallons of water per day. If a water-supply problem exists in your area, you might get in contact with the Washington Suburban Sanitary Commission, 4017 Hamilton Street, Hyattsville, Maryland 20781.

Even if you don't consider the direct or indirect costs of water in your home energy-conservation budget, the higher prices and diminishing supplies will catch up with you sooner or later. One way to analyze your water consumption is to make contact with your second meter. Furthermore, the water meter can be read and used much as your electric meter. It can help you detect leaks, and it can tell you how much water your various appliances use.

Water meters come in two types, in the same way as electric meters. Some have clock-like dials that are read just as the electric-meter dials described in Section 2 of this book. Others have simple numbers. Water meters record water consumption either in gallons or in cubic feet. If yours is calibrated in cubic feet, you can translate to gallons by multiplying the number of cubic feet by 7.5.

You can use your water meter in several ways, as described in a booklet published by the American Water Works Association, Denver, Colorado:

(1) *To keep up with conservation efforts.* If you

have been trying to save water and wonder if you have succeeded, read the meter every week, or subtract the reading on your last month's bill from the reading on this month's. That will give you the amount of water consumed over that period.

(2) *To quantify a water-related activity.* If you wonder how much water you use for sprinkling the lawn, just turn on the sprinklers, watch the meter dial move around, and time it for a minute. Multiply the number of gallons by 60 for the quantity of water used per hour. You can also use the meter to figure out how much water you use for dishwashing or clothes washing, or showers and baths, by reading the meter before and after each of these activities.

(3) *Leak Detection.* Turn off all water taps in the house and read your meter. After a period of 30 minutes to an hour, read it again. If the dial hasn't moved, you have a water-tight home. If it has, you should check hose connections, faucets, and toilets.

(4) *Cheerleading.* The meter gives you daily feedback on your conservation efforts.

A Summary of No-Cost Quick Fixes, Money-Saving Approaches, and Maintenance for Your Energy Systems

HOT WATER   **Possible Annual Savings: $30 to $120**

Quick Fix:
Turn down the thermostat.

The Money-Saving Approach:
(1) *Time your heater.*
(2) *Reduce demand:* restrict the number of times (and length of time) you (A) wash dishes, (B) wash clothes, (C) take baths or showers
(3) *Do full loads of dishes.*
(4) *Cooler-wash the clothes.*
(5) *Use the dryer less.*

Maintenance:

(1) *Drain the tank.*
(2) *Stop faucet leaks.*

HEATING
Possible Annual Savings: $60 to $200

The Money-Saving Approach:
(1) *Lower the thermostat.*
(2) *The night setback.*
(3) *Zoned heating.*
(4) *Shut the vents.*
(5) *Close the drapes.*
(6) *Bring on the fans.*
(7) *Humidify.*
(8) *Use* made *heat.*
(9) *Activate the people.*
(10) *The oasis concept.*

Maintenance:
Check the furnace.

COOLING OR AIR CONDITIONING
Possible Annual Savings: $120 to $500

Quick Fix:
Turn off the sump heater.

The Money-Saving Approach:
(1) *Increase the air-conditioner thermostat setting.*
(2) *Use drapes.*
(3) *Close vents.*
(4) *Bring on the fans.*
(5) *Lay off the heat.*
(6) *The dehumidifier principle.*

Maintenance:
The yearly checkup.

Shedding the Load

COOKING AND THE KITCHEN
Possible Annual Savings: $20 to $130*

Quick Fix:
Douse the pilots.

The Money-Saving Approach:
(1) *Don't preheat the oven.*
(2) *Use the correct burner.*
(3) *Don't peek.*
(4) *Coast the roast.*
(5) *Use the right pan.*
(6) *Boil less water.*
(7) *Use the thermos.*
(8) *The one-pot principle.*
(9) *The wedding-present principle.*
(10) *Cold-dish meals.*

Maintenance:
(1) *Keep reflectors clean.*
(2) *Clean the oven after cooking.*

REFRIGERATOR/FREEZER AND DEEP FREEZER

Quick Fix:
Unplugging the oldies.

The Money-Saving Approach:
(1) *Stop the stare.*
(2) *Defrost it.*
(3) *Let it cool.*
(4) *Don't overload.*
(5) *Place it well.*
(6) *Don't sweat it.*
(7) *Turn it off.*

Maintenance:
(1) *Clean the coils.*
(2) *The dollar-bill test.*

---

*This includes savings from your refrigerator/freezer and deep freezer.

LIGHTING Possible Annual Savings: $10 to $40

Quick Fixes:
    (1) *Cut the wattage.*
    (2) *Take one out.*

The Money-Saving Approach:
    (1) *The automatic turnoff.*
    (2) *Turn out constantly burning, unnecessary lights.*
    (3) *Concentrate the light.*
    (4) *Use fluorescents.*
    (5) *How to buy bulbs.*
    (6) *Outside the house.*

ENTERTAINMENT (AND MISCELLANEOUS ITEMS)
Possible Annual Savings: $25 to $50

Quick Fixes:
    (1) *Check the swimming-pool pump.*
    (2) *Eliminate the Instant-On.*

The Money-Saving Approach:
Use a black-and-white TV instead of color.

SECTION 4

Energy Gadgets: Where the Payback Exceeds the Price

Let's take a second run through the house, this time with an eye out for buying various attachments and improvements that are now being sold to help you save energy. Many of these items take the place of the human exertions required in the last section. A thermostat timer, for instance, can be a welcome substitute for the cold-morning trek to turn up the thermostat. As the June 1976 Princeton energy report suggest, "A clock timer never forgets"; so if your voluntary conservation efforts are flagging, you might consider some technological help. If your family has resisted the plan for taking shorter showers, you can get the same energy savings, and keep pleasing them, by installing a flow-restrictor showerhead. If you have forgotten to put your water heater on a schedule, a bit of insulation will accomplish the same result. In other words, if you haven't achieved the 25-percent reduction in your bill(s), there are plenty of gadgets that will help you do so.

Even if your *no-cost* energy plan seems to be succeeding, you can increase the savings with some of the items described in this section. By carefully choosing among the items, you may get up to a total savings of 25 to 35 percent off your current utility bills. Small portable heaters are sold that can help you zone your heating; devices, as well, that make the fireplace work efficiently; and fans, awnings, shades, and shutters that cut air-conditioning costs. Many of the items we will list and describe cost less than $20, so you might buy one or two with the money you have already saved!

Furthermore, most of the products described in this section are not supposed to pay back in seven years, or some other disturbingly future time, but within eighteen months of purchase! We can't *guarantee* such a quick return, but we have tried to organize this section so you learn about the cheaper alternatives first, the expensive ones later. It makes sense to review these less expensive items before considering things like solar water heaters, for example. If your overall hot-water consumption can be reduced by some of the attachments presented in this section, you may eventually require only a smaller and therefore less expensive solar installation (see Section 8).

We have tried, as much as possible, to estimate how much you could save by purchasing the following services or products. The benefit you get from each of the items, of course, is directly related to how much you use the appliance that will be affected by it. Before you buy anything, check your electric-power log or energy audit; and try to spend your money on aids for the appliances you operate the most frequently, to maximize your return.

After you install one of the gadgets or attachments, a careful series of readings of your electric meter, or a look at your gas or fuel-oil bills, should help you detect whether the expected returns are being realized.

Hot Water

As we mentioned in Section 3, the three major losses in the hot-water system occur at the tank itself, in the pipes that distribute the water to various places in the house, and in the extravagant use of hot water by people who don't know its price. Devises on the market limit all three of these losses. One of them even cuts the hot-water bill for long showers without threatening the pleasure of the lingering shower taker.

The advantage of modifying your water system first is that you get a high return for items that are not expensive, and that are easy to install. The following three projects, taken together, can reduce your hot water bill by 40 percent.

(1) *Insulate the tank.* One of the most worthwhile purchases you can make, by all accounts, is insulation for the hot-water tank. Most tanks now in service do not have enough insulation in their lining. They may even feel hot to the touch—which means that heat you have already paid for is escaping through the walls of the tank. The colder the air outside the tank, the more heat escapes.

The Federal Energy Administration says that wrapping a hot-water tank with insulation will save between $5 and $20 a year. This estimate agrees with the results of research done at the National Bureau of Standards (NBS). NBS concluded that water heaters can be made 8-percent more efficient with additional insulation. If you pay $250 a year for heating water, then an 8-percent savings will give back $20. A third analysis of water heaters predicts that the extra insulation will save 400 KWH of electricity or 3,600 cubic feet of gas. At 5 cents per KWH, this amounts to $20. Gas prices fluctuate from place to place, but in general terms insulating a gas hot-water tank will not be as lucrative an investment as insulating an electric one.

Insulating a water tank is easy. You buy some aluminum-backed fiberglass insulation—the thicker the better—wrap it around the tank, and tie it in place with string or tape. The materials should not cost more than $10. There is one safety precaution that must be taken with gas water heaters: Make sure the *air vents* are not covered by the insulation. Covering the vents would interfere with the combustion process. (There is no such vent on an electric water-heater tank.) If you are unsure about where the vents on the gas-heated tank are located, ask your appliance-maintenance people to install the insulation for you.

If you don't want to prepare the insulation yourself, there is a hot-water kit that you can now buy at many hardware stores; it is manufactured by Johns-Manville. The kit includes white vinyl-clad fiberglass insulation, tape to hold it in place, and detailed instructions. With the kit, wrapping the tank with 1½-inch insulation is no more difficult than putting on an overcoat. The kit costs about $20, or double what you pay if you buy the insulation and tape separately yourself. If you cannot find the kit locally, it can be purchased by mail from Solar Usage Now, Box 306, Bascom, Ohio 44809, for $19.95 plus shipping charges.

Insulating the tank is an investment that should pay off in one year—and may pay off in a few months. An insulated tank makes all the no-cost hot-water conservation efforts more rewarding: you get a much greater return, for example, by taking shorter showers, or by doing fewer loads of dishes, if the hot water you save isn't dissipated in a poorly insulated tank.

This relatively inexpensive modification will also diminish the usefulness of a clock-timer device that turns on the heater for short intervals during the day. It is possible to save money with such a timer, or, as discussed earlier, by manually putting the water heater on a three-hour schedule. We think insulation is a cheaper and less-complicated way to get the same result. With proper insulation, the heat losses in the tank will be so diminished that putting the heater on a timer wouldn't be worth the money.

(2) *Insulate the pipes.* In some homes, there may be as much as sixty feet of 3/4-inch piping between the water tank and a distant faucet. That length of pipe will hold 2 gallons of water, which have probably cooled off in the pipe since the last time you used the faucet. When you use the faucet again, you must run off that 2 gallons, plus an additional gallon of water to warm up the pipe, before any really heated water actually reaches your hands. If the hot-water faucet is used ten times a day, then you waste 30 gallons of hot water a day—enough to take one long, steamy bath. In a month you waste 900 gallons.

By insulating the pipes, you can eliminate about two-thirds of the losses, or 600 gallons a month. In the North, it costs about $7 to heat that amount of water. The insulation for 3/4-inch tubing costs about

65 cents a foot, and it is easy to install. For a sixty-foot pipe run, it would cost $39 to do the job. In the example described above, such a project would pay for itself in less than six months. After that, you would profit by at least $7 a month, and you would also no longer have to put up with the nuisance of waiting thirty seconds or so for the hot water to start coming out of the faucet.

(3) *Instant flow restrictors.* If your family doesn't react favorably to the notion of shorter showers, you can accomplish the same result with a water-saving device. With a flow restrictor, you can take indulgent showers and still come out like a conservationist.

Flow restrictors are easily installed at the showerhead. They reduce the amount of water that comes out of the shower, without sacrificing the hard spray that a lot of bathers crave. One such device, called Nova, and manufactured by Ecological Water Products, Inc., provides an average flow of 2.1 gallons a minute, while a regular showerhead releases from 8 to 10 gallons a minute. A ten-minute shower with a Nova flow restrictor will, according to the manufacturer's estimate, save 59 gallons over the same shower with a standard showerhead. The company suggests that it takes about 13 cents in fuel to heat up enough water needed for the average ten-minute shower, while a Nova shower of similar duration would require only 3.5 cents worth of fuel. With Nova, you save about 10 cents on every shower.

The Federal Energy Adminstration offers the prediction that a flow restrictor will save a family $24 a year in energy costs. That's in addition to the thousands of gallons of water that won't go down the shower drain. Even at $24 a year, the flow restrictor is a good buy. The Nova showerhead can be purchased for $16.95 from Ecological Water Products, 142 Spring St., Newport, R.I. 02840. If the estimates are correct, you get your money back in less than a year.

We single out the Nova showerhead only because the company that makes it provides detailed information on its performance. Other manufacturers, such as

American Standard, Dole, Crane, Kohler, and Moen, also sell flow restrictors and report similar results. Check the local plumbing-supply store to see what is available in your area.

These same companies also make aerators for bathroom and kitchen faucets. Aerators cut the standard water flow by 60 percent, while still providing plenty of pressure to blast grime and food off dishes. Ecological Water Products sells the aerator for $1.95, and estimates that it will save a family of four between 1,000 and 1,500 gallons of water a year. The aerator won't cut the hot-water bill as much as the shower-flow restrictor, but it will pay for itself quickly and provide a modest return thereafter.

(4) *Install a check valve.* This adaptation isn't one of the big three, but it can be profitable in certain instances.

Many hot-water systems lose heat through unwanted circulation. Small leaks in "mixing faucets," and in appliances connected to both the hot- and cold-water lines, can permit a constant flow of hot water through the entire tubing system of the house. If uninsulated pipes happen to cross the cold attic or the crawl space, then a lot of heat can be lost to the cold air during a continuous, hidden circulation cycle.

You can test your own water system to see if you have this problem. (For the test to be valid, it must be done early in the morning, before any hot water has been drawn.) Find the hot water line that comes out of the top of your water heater, and run your hand along the pipe, for as far as you can reach. If the pipe is still hot several feet from the tank, then unwanted circulation is probably the cause. You can partially eliminate that circulation by turning off the hot-water shutoff valve on the top of the tank before going to bed each night, and turning it on again in the morning. But a better solution is to have your plumber install a spring-loaded check valve in the direction of flow of the pipe coming out of the top of the tank. The valve costs about $7. A small spring holds the valve shut against the pressure of the unwanted circulation, but opens the valve whenever hot water is required by a person or by an appliance.

(5) *Eliminate the pilot light.* The National Bureau of Standards estimates that gas hot-water heaters could be made 9-percent more efficient if the pilot light were eliminated. You can have the pilot replaced with a solid-state ignition system for $38, plus installation; but this purchase is of dubious economic value. It would take a long time to make back your investment.

(6) *Buy a solar dryer.* An investment in a solar dryer (clothes line) is guaranteed to pay for itself within a matter of weeks. (Or, if you have a gas clothes dryer, you can eliminate the pilot light—which takes up about 15 percent of the cost of running the machine. The pilot can be replaced by solid-state ignition, through the same process we described above. It is estimated that the pilot uses 4,800 cubic feet of gas per year. At current regulated prices, the conversion to solid-state would save you $20 per year. If gas prices are de-regulated, the conversion will become much more profitable.)

Heating

(1) *Buy a timer thermostat or attachable timer.* The single most profitable purchase for your heating system, if you don't want to bother to turn the heater down at night, is an automatically timed thermostat. Such a thermostat will unfailingly produce the night setback, and return the house temperature to the comfort zone in the morning. Neither timer thermostat nor an attachable timer is cheap, but they do give a high return in a short time.

The Energy Research and Development Administration recently put out a chart to show the savings a night setback can produce in various parts of the country; (Table 7) find the nearest location to your home on the table. Using the cost of last year's heating bill, you will be able to calculate the advantage *you* can derive from setback.

Table 7: Approximate Percentage of Fuel Savings Across the Nation with an Eight-Hour Night Setback of the Thermostat

City	5° setback	10° setback
Atlanta	11%	15%
Boston	7	11
Buffalo	6	10
Chicago	7	11
Cincinnati	8	12
Cleveland	9	12
Dallas	11	15
Denver	7	11
Des Moines	7	11
Detroit	7	11
Kansas City, Mo.	8	12
Los Angeles	12	16
Louisville	9	13
Madison	5	9
Miami	12	18
Milwaukee	6	10
Minneapolis	5	9
New York	8	12
Omaha	7	11
Philadelphia	8	12
Pittsburgh	7	11
Portland, Ore.	9	13
Salt Lake City	7	11
San Francisco	10	14
Seattle	8	12
St. Louis	8	12
Syracuse	7	11
Washington, D.C.	9	13

You may notice in the table that the highest percentages of fuel savings are predicted for the warmer regions of the country. In cold areas—where the furnace stays on constantly just to produce livable indoor conditions—the setback isn't as effective in reducing the percentage of fuel consumed. Since people in cold climates spend much more money annually on heating their homes, however, they stand to make more money off the night setback than do people in warm climates. The average savings from a night setback, according to ERDA, vary from 5 to 18 percent. If your annual fuel bill is $750, that would mean a saving of between $37.50 and $135.

If you are wondering how such estimates might hold up in specific situations, Princeton University has had impressive results with automatically controlled thermostats in their residential test community. The thermostat used there, which is not yet available on the market, saved 18 percent on fuel bills.

There are two categories of thermostats that can be used to time your heating system. The first are the timer devices that can be added to an existing thermostat; they are easy to install and cost about $50. The second are the actual replacement thermostats; these can provide more flexibility than a simple timer, but they cost from $75 to $100, including installation. Some replacement thermostats offer multiple setbacks, through which you can maintain your house at two or three different temperatures during the course of the day or night, without resetting. Check with your local heating-appliance dealer to see what is available in your area.

(2) *Have your thermostat adjusted and cleaned.* If you haven't done it in a few years, it may be worth the money to have your existing thermostat adjusted and cleaned. The Princeton energy researchers were surprised to find that half the thermostats in the Twin Rivers community were inaccurate by a degree or more. According to the Princeton report, a one-degree error can add 4 percent to a fuel bill. (It can only cost money, of course, if you habitually set your thermostat to the same number.)

To operate properly, a thermostat must be absolutely level, and its internal workings must be calibrated. Find out how much it costs to clean and tune your thermostat. It may be a worthwhile addition to your budget for furnace maintenance.

(3) *Zone your heating.* The plan to limit the home comfort zones to a smaller number of rooms, as de-

scribed in the preceding section, can be helped along by using various types of portable heaters and removable insulation panels. Some families have already begun the money-saving process of turning the central heater down to about 55°, and using auxiliary heaters to raise the temperature to comfortable levels in the occupied rooms.

Many small, inexpensive heaters on the market don't use much fuel or electricity. The portable 1,500-watt electric space heaters with built-in fans sell for about $25 in any hardware store. Portable kerosene or fuel-oil heaters are also available.

Another way to heat a small area in a house is with a solar window. All windows are solar, in the sense that they let in sunshine. But two inventors named Bill Rankin and David Wilson discovered that by building a box outside a window to trap more sunlight, they could produce a flow of air as hot as 120° from the box. It doesn't take much 120° air to take a big load off a furnace.

A solar window is a perfect beginning project for a person interested in the principles of solar power but not ready to build a full-fledged collector system. The window consists of a glass-covered, insulated box that is attached to a standard windowsill. Plans for such a window can be ordered from Rankin and Wilson at Lorien House, P.O. Box 1112, Black Mountain, North Carolina 28711. A plan for a similar window can be found in a book called *Fuel Savers;* it can be purchased for $2.75 from Total Environmental Action, Harrisville, New Hampshire.

(4) *Use interior insulation.* Various types of insulating panels can be used to close off unoccupied rooms, or to seal furnace ducts in areas where no heat is required. *Mechanix Illustrated,* in its November 1976 issue, described one such panel, called Panelfoam, that only costs about $4 for a six-sheet package. The sheets are a little over thirteen inches wide and forty-seven inches long, and can be easily cut and installed in various parts of the house. The article, by Thomas Jones, suggests that the panels can be used to insulate the back of kitchen cabinets that are attached

to outside walls; to cover air conditioners to reduce cold-air entry; to fit into window frames as temporary storm-window substitutes; to seal edges around attic doors; and to cover unnecessary heater registers.

Other types of panel material can be used to make insulating room dividers in order to further reduce the size of areas to be heated. If you have large windows on a cold, shaded side of the house, you can also cut your fuel bill by blocking off some or all of those windows with insulating panels. Such panels are especially effective on the northern exposure, where very little sunlight gets through. You can make insulation panels out of cork, plywood, Celotex, or certain types of plastic. Some of the plastics, like styrene foam, give off toxic gases when they are ignited, however; so you should check the safety requirements. To install insulating panels, you cut to the size of the window and insert the panel right up against the glass. A description of how to make such panels is given in *Low-Cost, Energy-Efficient Shelter,* available at $5.95 from Rodale Press, Emmaus, Pennsylvania 18049.

(5) *Insulate the ducts.* If your furnace is located in an uninsulated basement, then insulating the furnace ducts should be a high-priority investment. We discuss most kinds of insulation in Section 6, since large home-insulation projects cost more than the items that are included in this section. But duct insulation is not prohibitively expensive, and installing it is an easy job. The cost of insulating furnace ducts in an average-sized basement with two-inch fiberglass, backed with aluminum, will run somewhere around $50. If you don't install it yourself, add $60 for labor. The payback should occur sometime between one and two years. In the townhouses studied by Princeton, the estimated payback time for duct insulation was between thirteen and thirty-two months, depending on whether the people did the work themselves or hired a contractor to do it.

(6) *Resize your furnace.* Most houses have much bigger furnaces than they need. Running an oversized heater or furnace wastes money. There is no way to crush your heater down to a compact size, but if you

have an oil burner or a gas burner, there are ways to reduce both its burning surface and its fuel requirements. Various companies now make devices that limit the amount of fuel that flows into the furnace, and also change the size of the burner, or orifice.

Unless you are a heating expert, you should not mess with your furnace's innards. Get your service man to advise you about the costs and benefits of reducing the orifice size and fuel flow. One company that makes furnace-conversion kits is Lennox Industries, Inc., P.O. Box 250, Marshalltown, Iowa 50158.

(7) *Buy a flue damper*. Several companies manufacture devices that automatically close the flue of a gas furnace when the burner turns off. This inhibits warm air from escaping up the chimney, and is supposed to save about 20 percent on total fuel consumption. The American Gas Association, 1515 Wilson Blvd., Arlington, Virginia 22209, has a list of such companies. You can write to the association for the list, or check with local sources to see if they carry flue dampers.

(8) *Get something better than drapes*. Most drapes or curtains have only limited value as a thermal barrier. In the winter, if furnace air can flow through the space between the curtain and the window, then the effectiveness of the curtain is limited. You can, of course, build boxes—valances—across the top of your draperies in order to seal the curtained area more tightly. Also, some companies make thermal drapes; and the Duracote Company even makes insulating drapes that you can seal to the sides of the window.

But you might also consider a variety of roll shades, which do a better insulating job. These shades are made of clear or coated transparent-plastic film, and are sold either in single layers or in multiple layers, with an air space between each piece of film. Some roll shades can be sealed to the sides of the window, thus providing a dead air space that further increases their insulating power. According to a booklet published by the National Bureau of Standards, a reflective film shade mounted on a roller costs about $2.15 per square foot, or about $32.25 for each 3 × 5 foot window. The 3M Company, Minneapolis, Minnesota, manufactures a low-emission film shade with magnetic tape on the edges for sealing it to a window jamb or -sill. Tests were done at a Minneapolis school in 1974 that show the insulating effectiveness of various materials as follows. The results were published in a book, "Window Design Strategies to Conserve Energy," put out by The National Bureau of Standards.

Effectiveness of Various Window-Insulating Materials	Heat loss reduced (relative to single-glaze window)
(1) Conventional roll shade	28–36%
(2) Clear-plastic film shade with all perimeters sealed	36–43%
(3) Wood frame exterior storm windows	50–57%
(4) Low heat-emitting film shade sealed at bottom and sides only (3M shade)	57–64%

The 3M shade was judged to be slightly superior to the storm window in reducing heat loss. It is about the same price as a storm window, but has the advantage of being installed *inside* the house, where it can be rolled up or down as needed.

Another company, Ark-tic-Seal Systems, Inc., of Butler, Wisconsin, makes a three-shade system, composed of a reflective shade near the window, a heat-absorbing shade near the room, and a clear shade in the middle. The shade system is described in the book mentioned above. The shades are operated in a frame, so that the sides and bottom are sealed. They can be lowered or raised in different combinations to get the most advantage of sunlight in winter without losing the furnace heat to the window. They can also be used in summer to keep out sunlight and to improve the efficiency of the air conditioner. You can write the company for further details.

(9) *Improve your fireplace.* As we mentioned earlier, fireplaces may take more hot air out of a house than they put into it. The furnace heat lost up the

chimney is often greater than the extra heat produced by the fire itself. We know two ways to get around this problem. One is to install glass doors on the front of the fireplace, and keep the doors closed. Very little heat from the fire will get into the room, but at least the furnace heat won't escape up the chimney. Many companies make such doors. The second method is to buy one of a number of heat-catching fireplace grates that circulate air or water around the fire, capturing more heat before it is lost up the chimney. The Heat-Catcher, made by Lassy Tools, Inc., Plainville, Connecticut 06062, uses a heavy iron tubular grate. A blower in the heat catcher circulates room air around the fire and hot coals. It works like a warm-air furnace, releasing a blast of hot air to the room. Energy-saving fireplace attachments like heat grates are now sold in large chain stores, such as Sears Roebuck.

(10) *Do energy landscaping.* Some of the best opportunities to save energy are found outside the house. We describe in Section 12 how trees, shrubbery, and windbreaks have reduced energy costs in certain homes by rather hefty percentages. A landscaping plan that takes energy into account can pay for itself in lower fuel bills.

The idea is, you want to find things that will stop wind during the winter but will not cut off the breezes in the summer. Deciduous trees are perfect summer energy savers, since they provide shade during the hot months and lose their leaves during winter, when you want sunlight to fall on the house. Shrubs are better winter energy savers, since shrubs can hold off the wind better than trees. Plant shrubs on the cold north side of your house. The most effective place to set shrubs or a windbreak is upwind, at a distance from the house that equals one and one-half to two and one-half times the height of the building. Since winter winds come from a different direction than summer winds, such shrubs or windbreak probably won't stop your summer breezes. A windbreak made of wood will work better if it is not solid, and if it provides space for *some* of the wind to get through.

Cooling or Air Conditioning

(1) *Invest in an attic vent or fan.* Energy researchers have only recently discovered the extreme usefulness of the attic exhaust fan as a fuel-saving device. In terms of return per dollar of investment, it is right up there with the thermostat timer. In warm parts of the country, roof temperatures climb to 165° in the summer, and the attic area becomes a veritable oven, putting tremendous strain on the air-conditioning system. If your attic is unbearably hot in the summer, and if it is not ventilated, an attic ventilator is a high-priority item. A good attic vent can even prevent the buildup of moisture in the attic during the *winter*—moisture that inhibits the effectiveness of insulation. In this way, an attic vent can help reduce heating costs, although it should only be opened periodically during the winter, so that heat doesn't routinely escape through the vent.

Many types of attic vent systems are available: fans that are installed in outside walls of the attic; soffit fans; revolving metal mushrooms that are mounted on the roof; etc. A good vent system should reduce attic air temperatures on hot clear days from 160° down to a more manageable 95°. It should have a thermostatically controlled fan that turns on at 85°, and should be large enough to change all the air in the attic every two or three minutes.

Attic vents are available in a variety of models and sizes. Check with your local building-supply store. The vents are usually not too expensive, but they must be installed correctly in order to avoid roof leaks. Get a vent with a damper that can be closed during the winter, so that valuable hot air is not lost when it is needed.

Another way to reduce air-conditioning costs is with an attic house fan, which is placed horizontally between the attic and the ceiling of the floor below. A university study that compared two homes in Houston, Texas, was described recently in *Popular Science*. One home had a large, horizontal attic fan, which pushed air through a well-ventilated attic; the other did not. In the house with the fan, the air-conditioner thermostat was set at 82°. In the house with no fan, the thermostat had to be set at 75° to produce equivalent comfort. The fan saved 5,300 KWH of electricity, or $265 in a year.

(2) *Buy a timer thermostat or attachable timer.* The automatically timed thermostat or timer attachment that saved money on heating bills can also be used to cut air-conditioning costs.

3) *Use drapes, shades, and sun-control film.* Solar heat gain is reduced considerably by simple draperies with a white backing, but there are some products on the market that do the job better. The multiple shades described earlier in this section not only add heat to the house in the winter, they also work to keep heat out in the summer.

For people whose summer cooling bills are larger than their winter heating bills, sun-control film might be a useful purchase. Sun-control film is a silvery plastic material with an adhesive backing that can be attached to the inside of any window. The material reflects up to 80 percent of the sun's rays, and reduces glare. It has the advantage of providing privacy during the day, as it is impossible to see through the film and into the house during daylight hours (though you can see out). This advantage must be weighed against the reverse effect at night; after dark, you can't see out, but those on the outside can see in. Sun-control film is not cheap. If you install it yourself, it costs about $1.25 a square foot. In areas that require several months of air conditioning, the film can be a good investment, but it is not an advisable purchase for cold climates, since it also keeps the sun's rays out of the house in the winter.

Several companies now make a removable sun screen. This looks like a normal, everyday screen, but it contains vinyl-coated fiberglass, and is supposed to reduce heat gain by 75 percent. The screen is installed on the outside of the window during the summer. In the winter, it can be rolled up. The cost is about $10 per window, and the screens can be ordered from the manufacturer, VIMCO Corp., Dept. WD, Box 212, Laurel, Virginia 23060. Another, similar screen is pulled down and retracted with a spring-operated roller. It is the Modern-Aire Solar Control Transparent Window Shade, made by Solar Usage Now, Box 306, Bascom, Ohio 44809. These removable sun screens come in many sizes, and cost from $14 to $45 each. Exterior sun screens are also marketed under the name Kool-Shade.

(4) *Purchase shutters, awnings, or exterior roll blinds.* Louvered shutters and awnings can shield the house from the sun's glare and stop a lot of heat from getting into the house. Awnings are made of various materials and designs. Check with your local outlets. They are most useful in warm climates, especially on the south, east, and west sides of the house. If you use air conditioners regularly and have unshaded windows on the hottest sides of the house, then you can probably benefit from awnings of some kind.

Shutters come in wood or metal. There are rolling shutters that move horizontally to cover windows, Bahama shutters that flap up or down, or hinged shutters that open across the window. A visit to the local supply store should fill you in on designs and prices. Light-colored shutters work better than dark ones. Shutters are also most effective if they don't trap the hot air around the window. Louvers provide a way to let the air flow through.

An exterior shading device can reduce solar heat gain through a window by up to 80 percent.

Exterior roll blinds also keep out the sun's heat. One manufacturer reports that air-conditioning costs can be reduced by 35 percent with such roll blinds. Not only do they work in summer, they can also be lowered during the winter to provide a dead air space that acts as insulation for the window. At night, the roll blinds can help keep heat in the house. Roll blinds have been used in Europe for many years and are found on 85 percent of European residential buildings and high-rises. In America, they are manufactured by the Pease Company, New Castle, Indiana 47362.

(5) *Shade your air conditioner.* A final modification for air conditioners is to make sure the compressor is properly shaded. It should be installed on the cooler, north side of the house, preferably under a tree or awning or other heat-reducing cover. If yours is out in the direct sunlight, you could improve its performance by building a lean-to or umbrella to shade the machine. But be sure that nothing blocks the free flow of air around the unit.

Cooking* and the Kitchen

(1) *Eliminate the pilot.* If you can't manually deactivate the stove pilot(s), or if you don't want to light matches continually, or run the risk of a gas leak, you can convert your gas stove to automatic electronic ignition. Conversion kits cost about $40, plus installation. The payback should occur in less than three years, and since gas prices are likely going up, the penalties for keeping the pilot lit are going to increase. Payback, in that case, will occur much sooner.

Check with gas-appliance dealers in your area to find out about conversion kits, and whether one can be installed on your stove.

(2) *Buy small appliances.* There is not much you can do to improve the efficiency of the electric oven or range you already own. But small appliances such as crock-pots, as discussed in the Section 3, use smaller amounts of energy. So do pressure cookers and woks. (If you want to buy somebody a cooking device, as a gift, consider buying something that will save on energy money!)

You may already have a few energy-saving kitchen devices, such as the meat thermometer. Many people count this as one of the most important conservation items in the entire house. A meat thermometer eliminates wasteful oven peeking, provides you with an

---
* Consideration of the refrigerator/freezer part of the kitchen system follows.

accurate timetable for food preparation, and does away with the costly practice of overcooking the meat. A second item is the thermos bottle. By using a thermos to keep the coffee warm, you don't have to keep one of the electric burners on constant Simmer. A third is a simple ceramic tile, that can be put in the oven while dinner is being cooked. Removed, the tile later takes the place of a bun warmer.

(3) *Purchase a newspaper grill.* Cooking meals outside in the summertime can be a good way to save on air-conditioning bills. One remedy for the high cost of barbecuing is the Shikari grill, which can barbecue steaks or pieces of chicken with four pieces of crumpled-up newspaper. During cooking, the fat drips down from the meat, saturates the newspaper, and is burned as fuel. An entire meal can be produced from one section of the local newspaper. We have tried it, and it works well. The grill, which looks like a metal garbage can, is light, easy to disassemble, and makes an excellent survival stove in case of a blackout or fuel shortage. It is available from Norm Thompson, 1805 NW Thurman, Portland, Oregon 97209, for less than $20.

(The ultimate energy saver for cooking is the solar cooker. Several varieties are available at camping equipment stores.)

Refrigerator and Freezer

(1) *Replace the gasket.* In an earlier section, we described the dollar-bill test of a refrigerator gasket: placing a bill between the door gasket and the metal edge of the box, closing the door, and pulling at the bill. If the bill can be easily removed, it's time to get a new gasket. They cost between $15 and $40, if you install them yourself, and are always a good investment if you need them. Be sure to have your appliance dealer show you how to install the gasket. It can be a little tricky, until you get the hang of it. If the old gasket was in really bad shape, the new one could pay for itself in less than two years.

(2) *Go back to manual.* The only other way to

revise your refrigerator is to "lobotomize" the automatic model so that it again requires a certain amount of human support; in effect, turn it back into manual. You can have your appliance person unhook the automatic defrost or frost-free mechanism, and save as much as 35 percent of electricity required to run the machine.

A detailed analysis of the cost of operating various kinds of non-manual refrigerators is found in the next section. But in general terms, the automatic defrost can cost you anywhere from $1 to $3 a month, at current prices. When electric rates double, you may be paying $75 a year for the privilege of frost-free refrigeration. If you are willing to settle for *partial* automatic defrost, you can have the appliance person install a single-pole, single-throw toggle switch (as described in the April 1977 issue of Popular Science). Such a switch, which costs about $1.25, will permit you to use the automatic defrost, but only when it is needed.

Also, while you are unhooking things, you can deactivate the door sweat heater and save from 120 to 300 KWH, or between $6 and $15 per year.

Lighting

(1) *Convert to fluorescent*. For light fixtures that are in constant operation, it makes economic sense to convert to fluorescent bulbs. A double-bulb incandescent fixture that uses two 100-watters for 10 hours a day will cost $36 in yearly electric bills. A fluorescent unit can provide a similar amount of light for $7.20. The $20 or so it costs to convert to fluorescent will be paid back in less than one year. But fluorescents, remember, should not be used where continual on-and-off switching is required. Constant switching shortens the life of a fluorescent tube.

If you don't want to buy an entire fluorescent fixture, there will soon be fluorescent bulbs that fit into regular incandescent sockets. These bulbs will cost about $7 each, but will last much longer than regular bulbs. In places where constant lighting is required, they will pay for themselves in a matter of months.

(2) *Time the outside lights.* If you have outside lights that burn all the time, it may save you money to attach them to a clock or photo-cell timer. Estimate the electricity cost for keeping such lights on, and compare this to the cost of the timers available.

Entertainment (and Miscellaneous Items)

Buy a solar pool cover. Some people reduce the heating bill for their swimming pool by covering the whole thing with a plastic cloth. A California company manufactures a specially designed plastic cover that floats on top of the pool; send for prices and information to Dearing Solar Energy Systems, 12324 Ventura Blvd., Studio City, California 91604. Other solar floating devices are sold through the Solar Usage Now catalogue. Your local pool manufacturer or solar pool heating specialist should also know where to obtain information on these materials.

Special: Saving Water

While it takes a little thought and planning to reduce an *electric* bill, it is much easier to cut *water* consumption by thousands of gallons a year. We have already discussed shower-flow restrictors that save on both water and electricity, and many other adaptors and gadgets are available that save on water alone.

One home remedy for the 5-gallon flush is to put a brick or a rock in the water-supply section of the toilet, thus reducing the amount of water that is available to flow down into the toilet bowl. However, bricks can damage the inner parts of the toilet or interfere with the flushing action, and really should not be used. Several inexpensive devices on the market do the same job with no attendant risks. They are described in an excellent free publication on the merits and ways of saving water put out by the Washington Suburban Sanitary Commission, 4017 Hamilton St., Hyattsville,

Maryland 20781. The devices are categorized as follows:

(A) *Toilet-valve assemblies*. These are toilet balls that guard against leaks in the toilet system. They make some sort of noise if the tank ball is defective. One is manufactured by Fluidmaster, P.O. Box 4264, 1800 Via Burton, Anaheim, California 92806. Other valve assemblies with the leak-detection feature are sold under the names Exelon, Control Guard, Mansfield Water-Saver, and American Standard.

(B) *Toilet-flush adaptors*. These gadgets permit a "light flush" for liquid waste and a "full flush" for solid waste. They eliminate one of the problems of the brick technique (which sometimes does not provide the necessary amount of downflow during the flush). They are available from Utah Marine, 459 S. Seventh St., P.O. Box 485, Brigham City, Utah 84302; from Ecology Helper, 106 W. Jefferson St., Falls Church, Virginia 22046; and from Gold Ring, Jefco Ave., Pacific Grove, California 93950.

(C) *Water-closet inserts*. These are substitutes for the brick. Various companies make them, and they range in cost from $3 to $6. Approximate water savings in most toilets will be about one to two gallons per flush. They are sold under various names, including: Aqua Guard, Littlejohn, Moby Dike, Aqua Miser, Water Wizard, and SA-720 Watersaver.

(D) *Water-saving toilets*. Most standard toilet manufacturers put out a water-saving toilet. We even describe some expensive and innovative new toilets that don't use water at all, in Section 5. But for people who are planning to install a traditional toilet, there is no reason not to buy the water saver.

Additional Energy-Saving Ideas

(1) *Buy a power meter or energy monitor to watch your watts.* If you get very involved in counting KWHs, the power meter makes a better measuring device than the electric meter. A power meter is plugged into the wall socket, and the appliance you want to measure is plugged into the meter. The meter registers the hourly

cost of running the appliance, in dollars and cents. It can be purchased, for about $20, from EM Electronics, P.O. Box 50367, Tucson, Arizona 85703.

A fancier way to keep up with your electricity bill is the Fitch Energy Monitor, put out by Fitch Industries, 203 N. Greensboro St., Carrboro North Carolina. The energy monitor can be set up in the kitchen, to give an instantaneous reading of how much power is being consumed in the house at any given time. It, too, is calibrated in dollars and cents—to avoid the translation from KWHs. The Fitch Monitor costs over $100, which takes it out of the gadget/gift range. The government is currently studying how effective these devices are in helping people reduce energy consumption.

(2) *Send for an energy catalogue.* There are not too many catalogues in which you can browse through the energy-saving-device section just as you browse through the appliance section in a Sears Roebuck catalogue. Manufacturers of these energy-saving items tend to be small companies that are dispersed all across the country.

We can report, however, on two attempts to list the various devices and gadgets in one place. The first is a mail-order company called Solar Usage Now. They publish a 127-page catalogue that describes all kinds of solar equipment, wood-burning stoves, and small energy-saving devices; it is available for $2 from the company at P.O. Box 306, Bascom, Ohio 44809. The second is the J. S. Young Company, which puts out a 32-page catalogue on environmental products. It includes sun-control film, attic fans, wood-burning stoves, solar thermostats, ceiling fans, and weather instruments. You can get a free copy by writing this company at 5621 East Calle de Paisano, Phoenix, Arizona 85018.

SECTION 5

The Hidden Dividend: How to Buy Large and Small Appliances and Systems

Just as cars use different amounts of gasoline to travel the same distance, appliances use different amounts of electricity to do the same work. Only the wealthiest among us still buy cars without showing some interest in the EPA mileage ratings, but it's a rare appliance buyer who gives even a moment's thought to energy efficiency ratings. This is not altogether the fault of the customer—because the salesmen don't exactly shout out the ratings, the minute you walk into the appliance store. In most cases, it takes a muckraker's disposition to even find out the energy rating of the appliance you want to buy.

These ratings are worth some attention. The people who make air conditioners, refrigerators, heaters, freezers, and stoves, etc., put out some kilowatt guzzlers and some kilowatt sippers; and you can't tell, by looking, which are which. One refrigerator may require 50-percent more electricity than another refrigerator of exactly the same size and with the same cooling power. The difference can add several hundred dollars to your utility bill, over the lifetime of the appliance.

Consider two 9,400-BTU room air conditioners, both made by York. (We single out York only because that company happens to make some very efficient air conditioners, along with some that have only average energy ratings.) One of them uses 860 watts; the other uses 1,365 watts. The 860-watt model undoubtedly has a better compressor and more cooling coils, and it costs $82 more to purchase than the 1,365-watter. Since they both put out the same amount of cool air, an economy-minded shopper would prefer to buy the cheaper model. But when you consider the cost of the electricity it takes to run the machines, the higher-priced model—paradoxically—ends up being the most economical.

In an area like New Orleans, air conditioners operate for an estimated one thousand hours a year. At 5 cents a KWH, the cheaper, 1,365-watt model will use $68.25 worth of electricity each year. The 860-watter will use only $43 worth of power. In four years, the more expensive air conditioner will make back the extra amount of money required to buy it. In thirteen years, it will save an additional $246.25 over the cheaper model—a sizable dividend for merely choosing one machine over the other. And if electric rates go to 10 cents a KWH, the thirteen-year saving on the more efficient machine will approach $492.

That is merely the savings you get from a single-room air conditioner. The differences in energy costs for a houseful of such appliances could pay the family car insurance, or send a kid to camp. Calculating those differences is part of a process called "life-cycle costing." It will eventually change the way careful consumers approach the appliance store.

Understanding Life-Cycle Costing

The idea behind life-cycle costing is that *the electric or gas or fuel-oil bill for each appliance or system should be considered as part of the cost of that appliance.* When you buy a house or a car, you consider the downpayment and the monthly interest payment. When you compute the life-cycle cost of an appliance, you are figuring the electric bill as part of the monthly interest payment. Buying one of the least-efficient appliances has the same budgetary effect as going to a loan shark.

Life-cycle is a newcomer to appliance calculations. People who peruse consumer magazines, hoping to save some maintenance money on a higher-quality product, are rarely aware of the greater dividends that can be achieved through life-cycle shopping. Nobody, of course, who has the information would turn down a superior product that also actually costs them less money in the long run, but the problem is getting the right information. A few companies now trumpet the virtues of their single most energy-efficient appliance, but they don't go out of the way to tell about the energy ratings of their less-efficient models. If people can't calculate the relative energy saving from one model to another, they will continue to prefer the products with the smaller price tags.

The Massachusetts Energy Office did a study of appliance buying in the Boston area, and concluded that people were just not buying the energy-saving refrigerators which, as we shall see, would save them a lot of money. It may be that salesmen don't push the energy-saving products, or that consumers are still more enamored of automatic ice makers than of lower utility bills. Or it may be that appliance buyers don't trust salesmen who promote higher-priced products that supposedly save money in the long run. Whatever the reason, the savings can be so substantial that we may only conclude that life-cycle costing is still not fully understood by the average consumer.

Currently, life-cycle costs are spelled out only for room air conditioners; the federal government now requires all manufacturers of room air conditioners to include an energy-efficiency ratio label on the machines. The government is planning, however, to mandate that all appliances carry such labels. But, knowing how long it takes the government to do things, we would be better off, now, to figure out our own life-cycle comparisons.

The following appliance and system descriptions contain our suggestions on how to buy air conditioners, refrigerators and freezers, heaters and furnaces (including heat pumps and wood stoves), stoves and ovens, hot-water heaters (including washing machines, dryers, and dishwashers—all of which rely on or relate to the water-heating system), televisions, and waterless toilets.

Air-Conditioners: The Money's in the EER

The first thing you want to know about buying an air conditioner is how big-powered a machine to get. The output of an air conditioner is expressed in BTUs, or British Thermal Units. The more BTUs of heat that an air conditioner can remove in an hour, the more powerful the machine. The tendency in the past has been to buy a much larger air conditioner than was necessary, but the price of electricity now makes such excesses highly undesirable. If you are buying a new air conditioner, or replacing an old one, consider the smallest size that will do the job.

One way to "size" an air conditioner is through the WHILE formula. We heard about it from the Consolidated Edison (Con Ed) utility in New York. "W" stands for the width of the room in feet; "H" is height; "I" is insulation; "L" is length of the room; and "E" is exposure. Once you figure WHILE, you divide it by 60 to get the BTUs you need.

Con Ed says to use 10 for the insulation factor in a well-insulated room, and 18 in a room that is poorly insulated or that has a lot of windows. The exposure factor is determined by the direction in which the longest outside wall faces. Con Ed suggests these figures: north—16; east—17; south—18; west—20. In a room of 15 × 10 × 20 feet, well-insulated, with the wall facing south, the formula would read:

$$\frac{\overset{W}{15} \times \overset{H}{10} \times \overset{I}{10} \times \overset{L}{20} \times \overset{E}{18}}{60} = 540,000 = 9,000 \text{ BTUs}$$

The power of the air conditioner, of course, can be directly related to how well your house or apartment has been insulated, weather-stripped, or otherwise modified for saving energy. When you are buying a new cooling (or heating) machine, you might consider comparing the cost of it to the total cost of a conservation package which would include insulation, attic fans, etc., and a perhaps *smaller* air conditioner (or heater).

We can't make any exact calculations, but here is what happened in Davis, California, when energy consultants were asked to prepare such an overall budget for a local family that wanted to install air conditioning. They figured that a 3-ton air conditioner would cost $2,500. Then they devised an alternate package that included $1,215 for insulation and weatherstripping, plus $300 for a lower-powered room air conditioner that would then be adequate. The total was $1,515. That's an initial saving of about $1,000 —plus continued savings on the utility bills.

The second thing to decide about air conditioners is whether you need room units or a central air system. A central system may produce an equal number of BTUs for less energy money than a room-by-room system, but room air conditioners have the advantage of flexibility. They don't all have to be turned on at once. Room air conditioners offer the possibility of zoned cooling. All things considered, we think room air conditioners are a more energy-efficient and economical choice.

The third thing to decide is which model or brand of air conditioner to buy. If you are in the market for a room air conditioner, don't overlook the EER numbers. They represent the output of the machine, expressed in BTU hours, divided by the electrical input into the machine, expressed in watts. The *higher* the EER, the *less* it will cost you to operate the air conditioner. A machine with an EER of 10 will run at half the cost of a machine with an EER of 5. You could have chosen correctly between the two air conditioners described at the beginning of this chapter by looking at their EER ratings. The less-efficient unit had an output of 9,400 BTUs, divided by its 1,365 watts—to produce an EER of about 6.9. The energy-saving machine had the same output—9,400 BTUs —but its wattage was only 860, which gives an EER of approximately 10.9. Anything over 10 is an excellent EER.

Unfortunately, the government does not yet require

EER labeling for central air conditioners—where the difference in ratings can add up to hundreds of dollars. As a matter of fact, a 24,000-BTU unit with an EER of 5.5, operating one thousand hours per year, will cost $786 more in electricity bills over a 10-year period than will a 24,000-BTU unit with an EER of 8.6. The higher-powered the air conditioner, the more critical the EER ratings become.

A little detective work on your part will literally save you a bundle of money. To calculate EER for central air conditioners, you need only know two things: the BTU output and the wattage. Both numbers should be printed somewhere on every air conditioner in the appliance store. If the numbers are not prominently displayed, ask the salesman where they are. Divide the BTUs by the watts, and you have your EER. Once you know the EER for various central air conditioners, then it's easy to make some rough calculations: an EER of 6 will produce 33-percent higher electric bills than an EER of 9; an EER of 5 will cost twice as much as an EER of 10. In addition, ask the salesman to help you figure out the *yearly operating cost* of each air conditioner that you might want to buy. You can compare those electric costs to the purchase price of the machines themselves, to get the optimal deal. In our opinion, there is no reason to buy any air conditioner with an EER lower than 8.

Refrigerators, Freezers, and Deep Freezers

The electric bill for two refrigerators of similar capacity can vary as much as it does for air conditioners. It depends on the amount of insulation in the boxes and the size and quality of the compressor. The government doesn't require any energy labeling on refrigerators and freezers, but fortunately the Association of Home Appliance Manufacturers (AHAM) publishes a booklet that lists the number of KWHs consumed in a month of operation by each refrigerator and freezer carrying the AHAM certification seal. Although the power-consumption data is furnished to

AHAM by the manufacturer, it is very accurate because it is subject to challenge procedure.

Going from the energy savers to the energy gluttons, the discrepancies in operating costs are enormous. Even at 4 cents per kilowatt hour, refrigerators of 13.5 cubic feet and above can use anywhere from $1.70 worth of electricity a month to $2.90 a month. For combination refrigerator/freezers that have between 13.5 and 16.5 cubic-foot capacity, the electric bills fluctuate from $2.60 per month for an efficient manual-defrost model, to $6.30 a month for an inefficient machine with completely automatic defrost. Depending on the quality of the machinery, the amount of insulation, and various added attractions, such as ice makers and automatic defrost, you could pay $600 in extra electric bills during the lifetime of one refrigerator, over another of exactly the same dimensions. At 8 cents or 10 cents a KWH, that extra amount could exceed $1,000.

The AHAM booklet, available for 50 cents from the association, at 20 North Wacker Dr., Chicago, Illinois 60606, is useful for the shopper who might want to save that $600 to $1,000. Without the booklet, it is very hard to compare the energy ratings of various refrigerators, and about the only thing you can do is to feel around the doors and walls until you find the refrigerator with the thickest insulation. The "feel" method, however, is far from foolproof, and the booklet will tell you everything you want to know. It lists the kilowatt ratings of many, but not all, popular refrigerators and freezers; and it is easy to compare the various models. Most manufacturers offer both high-efficiency and low-efficiency refrigerators of similar dimensions. Amana and a few other manufacturers now produce energy-saving models; these machines have more efficient compressors and better insulation. Amana reports that its 16-cubic-foot manual-defrosting energy saver uses about 1.7 KWH per day to operate. At 5 cents per KWH, it could run a full year on about $30. The Amana energy saver costs about $93 more than a conventional refrigerator of similar size that sells for $516.95. But, during the fifteen-year estimated life-cycle of the machines, the conventional model will run up a $1,080 electric bill, whereas the energy saver will cost only $459 to operate. Even if the price of power remains as it is, the savings on the Amana energy efficient model would be $528 ($621 savings in electricity minus $93 extra purchase price). If electric rates continue to rise at a rate of 10 percent a year, the savings will be well over $1,000. Either way, it is a good return on a $93-extra investment.

In buying a refrigerator, it is also worthwhile to consider whether the extra energy bills justify conveniences like the automatic defrost. The frost-free mechanism is one of the most notorious meter spinners of all the gadgets in the house. "Automatic defrost" is a euphemism for "heater," and any time you turn on a heater inside a cooling machine, there has to be a tremendous thermal collision. The cooler must work overtime to catch up—which makes automatic defrost a very costly baked Alaska. Since refrigerators are standardized, they are designed to work in the most humid and adverse conditions in the country. Automatic defrosters do so regularly and frequently, so that in less humid areas, where such numerous defrosts are unnecessary, the machine does them, anyway—and you pay for it. The AHAM booklet lists the various energy costs for machines with manual defrost that have between 13.5- and 16.5-cubic-foot capacity. Those machines can use between $2.70 and $5.40 each month in electricity. The same size boxes with automatic defrost cost from $4.00 to $6.80 a month to operate. For larger refrigerator/freezers, in the 19.5- to 22.5-cubic-foot range, the price you pay for automatic defrosting is even greater; it costs between $3.30 and $4.90 a month to run the manual defrost models, and from $6.00 to $8.00 a month to run the automatics. You may be shelling out an extra $36 a year for the pleasure of not defrosting your own machine, and, in the near future, that pleasure may cost $72 a year, or $1,000 over the lifetime of the refrigerator.

For people who reject the idea of defrosting their refrigerator, but who also want to save some energy money, Amana makes an energy-saving frost-free model. It uses about 2.8 KWHs of electricity a day. Its conventional counterpart uses up to 6 KWH a day. We can compare the costs as follows:

Comparison of Life-Cycle Costs of Frost-Free Conventional and Frost-Free Energy-Saving Refrigerators

	Frost-free conventional 16-ft. refrigerator	Frost-free Energy-saving 16-ft. refrigerator
Initial cost	$569.95	$664.95
Operating cost (based on 15 yrs. at 5¢ per KWH)	1,620.00	756.00
Life-cycle cost:	$2,189.95	$1,420.95

The anti-sweat device in some refrigerators also draws some extra electricity, but not as much as the automatic defrost. According to the AHAM list, the anti-sweat heaters add from 3 to 24 KWH a month; at 5 cents per KWH, you may be contributing $15 a year to the fight against icebox perspiration.

Deep freezers have varied appetites for electricity, just as do refrigerators. The AHAM booklet lists one 15.8-cubic-footer at 169 KWH a month, while another product of similar size uses only 68 KWH. These are extremes, but the differences in electric bills for similar-sized freezers can run as high as 100 KWH, or $5 a month; $60 a year; or $900 over the life of the machine. That's more money than a deep freezer costs in the first place.

If you aren't making an immediate purchase of a freezer or refrigerator, and like to hunt for bargains, then waiting for the AHAM booklet can be a very profitable delay. If you can't wait, then buy one of the energy savers. It will undoubtedly turn out to be a money saver as well.

Heaters and Furnaces

There are so many kinds of heaters that it would be impossible to evaluate all of them here; and many of them may or may not be available in your locality. But the general questions you want to ask yourself are: (1) What fuels are available? and (2) Which heaters produce the most heat for the least fuel cost?

By comparison with these two questions, the most frequently asked question—"Which heater is the cheapest to buy?"—is rendered less significant. Like air conditioners, refrigerators, and freezers, heaters use so much fuel that the cost of that fuel exceeds the purchase price of the heater. Indeed, the best way to think of heaters is that they are large boxes that burn money.

The proper sizing of a furnace is important to fuel economy. Chances are that yours is too big. Most contractors have sized home furnaces to meet the demands of the coldest temperature ever recorded in an area, plus 10 percent over, for extra measure. That's nice news for the once-in-a-century cold spell, but not so nice for the once-a-month fuel bill. If you are considering a new furnace, you might find out how your old one was sized, and see if you could do with a smaller unit.

Which ones burn the least "money"? This depends on several factors. The first is the *price of the fuel* itself. In most areas of the country, natural gas is the cheapest fuel, followed by fuel oil and then electricity. If natural gas is available, it may make sense to buy a heater that uses it. There are reasons to be cautious, however. Natural-gas prices are expected to take a precipitous jump soon, and natural gas may not always be easy to obtain. *Availability* is the second factor to consider. While gas and oil continue to be cheaper to buy, the price won't help much if you can't buy them at all. Many people reluctantly choose electric heaters because they feel that electricity is the only reliable source of power in their area. This may not always be true, either; but at least for now it seems that electric service is not plagued by as many problems of distribution and potential shortage as are natural gas or fuel oil. The *efficiency of the heating system* itself is the third factor. Electric heaters are 100-percent efficient, which means that all the electricity used by the heater can be converted into heat. Electricity, however, is not efficient to produce; over 70 percent of the energy is lost in the generating plant and power lines. Since you ultimately pay for that inefficiency, electric-resistance heating is a relatively costly proposition. Gas and oil, on the other hand, are not as efficient in the home furnace itself, but nothing is lost in the transmission and distribution. That's one of the reasons why gas and oil are cheaper fuels.

If your house requires only a heater, you might consider some of the wood-burners, or even the furnaces that combine wood and oil, as described in this section. If you need both heating and air conditioning, and you don't like the zoned-heating idea, then you might take a look at a heat pump. Heat pumps combine the two functions in a very economical way.

There is not too much good information about the relative cost of heating with various fuels. One such study, however, was done by Gordian Associates, and a chart from it was published in *Popular Science* magazine (Table 8). It compares costs for running a heat pump, a gas furnace with central air conditioning, an oil furnace with central air, an electric furnace with central air, and zoned heating with baseboard heaters and room air conditioners.

Gordian found that gas furnaces are very economical, a conclusion that is predictable. They did point out, nevertheless, that there are problems with natural-gas availability. Zoned heating, with small room units, is also economical, and supports the wisdom of the zone concepts advanced in our plan. For other types of heating systems, the heat pump proved more economical than oil furnaces or central electric furnaces in many of the cities studied. The Gordian study was done in 1975. Since that time, fuel costs have risen,

Table 8: How Heat Pumps Compare in Selected Cities

City		Period (months)	Heat Pump (kwh)	Gas furnace and central air conditioning Gas (cu. ft.)	Gas furnace and central air conditioning Elec. (kwh)	Oil furnace and central air conditioning Oil (gal.)	Oil furnace and central air conditioning Elec. (kwh)	Electric furnace and central air conditioning (kwh)	Baseboard resistance heaters and window air conditioning
HOUSTON 1290 degree days 1900 cooling hrs.	total heating season total cooling season total annual	4 8	1416 9283 10699	18137 — 18137	60 8019 8079	134 — 134	64 8019 8083	3372 8019 11391	3092 4818 7910
COST ($)			688.80	544.80		685.72		574.04	410.21
BIRMINGHAM 2483 degree days 1350 cooling hrs.	total heating season total cooling season total annual	6 6	3832 6790 10622	43224 — 43224	144 5867 6011	347 — 347	154 5867 6021	8107 5867 13974	7432 4455 11887
COST ($)			697.66	572.54		767.00		684.88	499.72
ATLANTA 2821 degree days 1000 cooling hrs.	total heating season total cooling season total annual	5 7	3837 6333 10170	43965 — 43965	147 5485 5632	340 — 340	157 5485 5642	8286 5485 13771	7596 4235 11831
COST ($)			700.78	560.57		743.77		656.74	482.98
TULSA 3670 degree days 1600 cooling hrs.	total heating season total cooling season total annual	6 6	7653 4731 12384	73358 — 73358	211 4365 4576	583 — 583	226 4365 4591	14364 4365 18729	13369 3501 16870
COST ($)			744.42	598.23		811.59		735.74	538.89
PHILADELPHIA 4508 degree days 700 cooling hrs.	total heating season total cooling season total annual	7 5	8013 4123 12136	80161 — 80161	279 3583 3862	610 — 610	299 3583 3882	15729 3583 19312	14386 2897 17283

COST ($)			885.89	743.09	935.28	944.61	728.23
SEATTLE 4407 degree days 100 cooling hrs.	total heating season total cooling season total annual	9 3	7682 1697 9379	76134 — 76134	579 — 579	14363 1382 15745	12818 1101 13919
COST ($)			477.29	518.30	684.70	473.23	321.83
COLUMBUS 5476 degree days 600 cooling hrs.	total heating season total cooling season total annual	7 5	10796 3896 14692	99738 — 99738	706 — 706	19199 3275 22474	17561 2804 20365
COST ($)			876.61	701.70	964.60	946.10	757.86
CLEVELAND 6097 degree days 500 cooling hrs.	total heating season total cooling season total annual	7 5	13242 2981 16223	116600 — 116600	826 — 826	22739 2597 25336	20798 2304 23102
COST ($)			943.29	637.59	954.83	1071.75	823.81
CONCORD (MASS.) 7377 degree days 400 cooling hrs.	total heating season total cooling season total annual	9 3	16631 1938 18569	130587 — 130587	948 — 948	26061 1813 27874	23784 1029 24813
COST ($)			1044.92	667.69	928.32	1219.93	1024.90

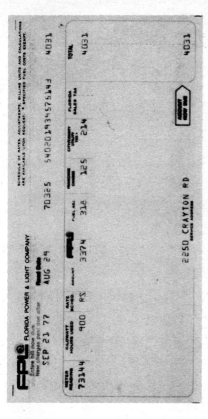

Plate 1. Sample Electric-Power Bill.
Courtesy Frank F. Tenney

This is an electric-power bill, similar to the one you receive from your power company each month. It tells you how much power (in KWHs) you have used in the past month, and how much money the power company is authorized to collect from you. To find out how much each unit of power costs, you simply divide the "Amount Now Due" by the number of "Kilowatt Hours used." In this case, $\frac{\$40.31}{900} = \$.0448$ per KWH. Your rate may be anywhere from 3 cents to 10 cents per KWH.

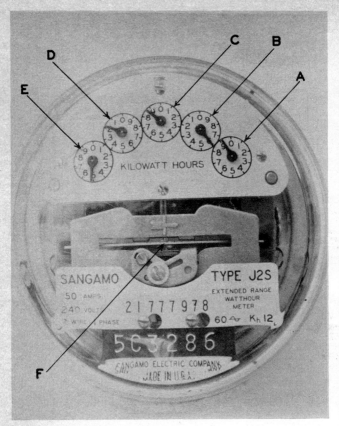

Plate 2. **Typical Electric-Power Meter.**
Photo by Frank F. Tenney

This typical electric power meter was installed outside the home of one of the authors in April 1964. When this photo was taken, on September 22, 1975, it was on its third time around, and read 51859 KWH. Before energy budgeting was started at this home in September 1975, the average yearly consumption was 19,884 KWH, over a five-year period. In the two years that energy budgeting has been used, power consumption has dropped to just over 56.7 percent of that figure — 11,280 KWH per year — proving that the system really works.

Month	Day	Present Reading	Power used last 24 hours	Remarks	Outside Temperature	Inside Temperature
Oct 1976	27	64415½	15½	Meter read late afternoon 64434	58	75
	28	444	28½	water pump 3½ hr. 9 KwH	61	75
	29	460	16		63	75
	30	476	16		61	75
*	31	502	26	water pump 3 hr. 7.5 KWH	69	75
Nov	1	527	25	Front passed Sunday afternoon 1hr water pump 2 Wash + Dryer Load	58	77
	2	541	14		63	77
	3	576	35	Cloudy Hot water heater on 6 AM off 9PM water pump 2½ hr 6½ KWH 12 KWH	59	77
	4	592	16		58	76
	5	608	16		58	74
	6	633½	25½	water pump 3½ hr 9 KwH	52	72
*	7	651	17½		63	73
	8	677	26	water pump 3½ hr 9 KwH	60	74
	9	696	19	2 Hr grill 3 KwH	49	70
	10	723	27	2 Wash + 2 Dryer Loads 9 KwH	59	72
	11	749	26	3¼ hr water pump 9½ KwH	63	74
	12	767	18	Very cloudy	66	76
	13	801	34	Hot water heater 18 KWH	69	78
*	14	839	38	Hot water heater 11 KWH water pump 9 KwH	70	75
	15	873	34	Hot water heater 5KWH 2 wash + 2 dry water pump 5 KwH Loads 9	73	79
	16	892	19	Hot water heater 5KWH To keep tank hot	66	78
	17	914	22	water pump 2 hr 5 KWH	70	79
	18	935	21	water pump 2 hr. 5 KwH	70	80
	19	951	16		65	78
	20	967½	16½		71	78
*	21	65002	34½	Water pump 8½ KWH Baked 12 lb. Turkey 9 KWH	72	79
	22	028	26	3 wash + 3 dryer loads 11 KWH	57	76
	23	051	23	Hot water heater 2½ hr 7 KwH	44	73
	24	068	17		50	72
	25	089	21	Space heater 4 hr 5 KwH (1320 w)	54	71
	26	130	41	water pump 3½ hr. 9 KwH Hot water heater 10 KwH	64	72
	27	146½	16½		72	76
*	28	175	28½	Hot water heater 11 KwH Grill 2 KwH	72	79
	29	206	31	water pump 3½ 9KWH 2 washer Meter read 65209 1 Dryer 6 KwH	75	80

Plate 3. Electric-Power Log.

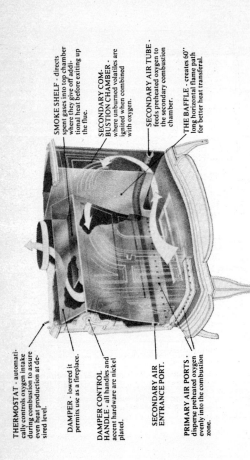

Plate 4. Defiant Parlor Stove.
Courtesy Vermont Casting Company

The cutaway of the Defiant Parlor Stove shows how complex a modern stove is; it gains high efficiency by secondary burning of the hot gases released by the burning wood.

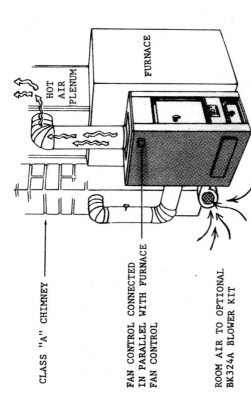

Plate 5. Diagram of an Add-A-Furnace.
Courtesy Malleable Iron Range Company

One solution to the problem of rising fuel costs is to install a Monarch wood-burning Add-A-Furnace. The stove pipe plugs into the existing class "A" chimney while the hot air passes through the existing duct system via the hot-air plenum.

Plate 6. Kitchen Heater and Stove.
Courtesy Malleable Iron Range Company

This attractive Monarch thirty-six-inch combination wood-burning kitchen heater with alternate electric elements is well adapted to cope with soaring fuel costs and shortages.

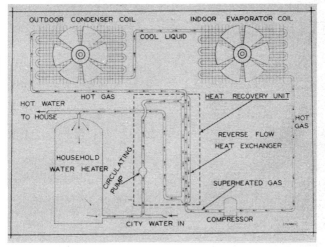

Plate 7. **Diagram of a Heat-Recovery Unit.**
Courtesy Frank F. Tenney

This diagram shows how the heat-recovery unit in the center — within the broken-line rectangle — captures waste heat from a central air conditioner and delivers it to the hot-water tank. This simple, relatively inexpensive device not only provides all the "free" hot water a family could possibly use, but in effect it improves the efficiency of the air conditioner, thus reducing its operating cost by up to 10 percent. This is almost a case of having your cake and eating it too.

Plate 8. Solar Home.
Courtesy Libbey-Owens-Ford Company

The look of the future. An artist's interpretation of a solar home, using Libbey-Owens-Ford Company's new Sun Panel Solar Collectors for both heating and cooling.

HOUSE SCHEMATIC
FOR DOMESTIC HOT WATER SOLAR SYSTEM

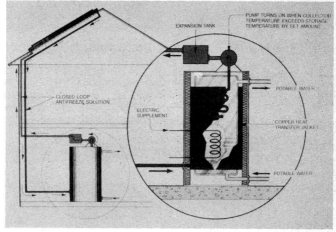

Plate 9.
Courtesy Libbey-Owens-Ford Company

A typical flat-plate solar collector system for heating domestic hot water. In this case, a closed-loop antifreeze solution heats the potable water by means of a heat exchanger. In a very warm climate, the potable water can be circulated through the collector, eliminating the need for a heat exchanger.

Plate 10. **Solar Collector Panels at NASA's Langley Research Center, Hampton, Virginia.**
Courtesy Chamberlain Manufacturing Company

One of the largest solar-energy research projects in the nation is underway at the National Aeronautics and Space Administration's (NASA's) Langley Research Center, Hampton, Virginia. In this test, over ten thousand square feet of Chamberlain solar collector panels are installed (as a major part of an array) to collect solar energy in the form of heated fluid. The energy thus captured will be used to provide most of the heating and cooling for the fifty-three-thousand-square-foot Systems Engineering Building seen behind the collector-panel installation. Every component in the system will be monitored for efficiency, reliability, and durability, as NASA joins in the search for cost-effective solar-energy systems.

Plate 11. **Flat-Plate Solar Collector Panel.**
Photo by Frank F. Tenney

This sixty-square-foot flat-plate collector was built from plans contained in the twenty-four-page booklet published by the Florida Conservation Foundation. The system cost about $1,000 to build, and has furnished a South Florida family with 99 percent of its domestic hot water during the summer and about 95 percent during the winter.

Plate 12. A Map of National Climate Zones.

Plate 13. Highland Avenue Branch of the Concord National Bank, Concord, New Hampshire.
Photo by Frank F. Tenney

The Highland Avenue Branch of the Concord National Bank was the first bank in New England to be both heated *and* cooled by solar energy. The twenty-six "Daystar 20" solar collectors shown here provide between 50 and 60 percent of the heating and cooling for this large structure. This is even more remarkable when you consider that the north, east, and west walls are completely glass, greatly increasing the heating and cooling losses. An Arkla absorption-cycle air-conditioning unit provides the cooling.

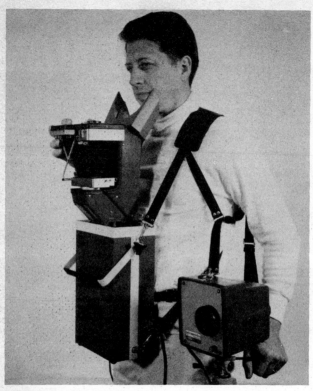

Plate 14. Portable Infrared Scanner System.
Courtesy Inframetrics, Inc.

This is a portable infrared scanner system, used to make thermographs to show heat loss from buildings. It was the type of equipment used on the "Red Rover" van by the Commonwealth of Massachusetts Energy Policy Office during Project Conserve.

Plate 15. **Reproduction of a Thermograph.**
Courtesy Inframetrics, Inc.

A reproduction of a thermograph made on a heat-loss survey of the Given/Rowell Medical Building Complex at the University of Vermont. The television-like presentation shows the temperature differences by the level of brightness in the picture. Black shows cold areas; the light areas show where heat is escaping from the building.

Plate 16. Energy-Saving Home.
Photo by Frank F. Tenney

The energy-saving residence of Evan Powell, a writer for *Popular Science* magazine. By carefully choosing his site, properly insulating the structure, and using zonal heating and cooling, Powell is able to heat and cool his home for about 25 percent of the cost of heating and cooling similar homes in the area. He has recently added a survival room to the house — which is fully described in the November 1977 issue of *Popular Science*.

and the exact numerical values would now be higher. But the thing to remember is that heat pumps have improved in efficiency since the Gordian study; if anything, they are more economical now.

Heat Pumps

Heat pumps got a bad name when some of the earlier varieties did not work well. However, since the late 1960s, many of the problems have been solved, and some good, high-efficiency heat pumps are now on the market.

A heat pump is a kind of reverse-cycle air conditioner. During the summer, it works like any other air conditioner, cooling air for the house and discharging hot air into the outdoors. In the winter, the heat pump simply reverses the process: it removes heat from the outside air and blows it into the house. Yes, there is plenty of heat in the air, even in wintertime!

Heat pumps cost more money to buy than air conditioners and heaters combined, but they require much less electricity to operate. A regular electric-resistance heater, we have seen, converts 100 percent of its energy into heat—for a coefficient of performance (COP) of 1. Heat pumps *extract* heat from the air, and they can achieve a coefficient of performance of 2 or even 3. This means that they produce twice or three times the heat for the same amount of electricity. They are, therefore, two or three times as economical to operate in the winter.

According to an article in the September 1976 issue of *Popular Science,* heat pumps tend to use a little more electricity in the air-conditioning phase than most standard air conditioners. But they more than make up for it in the winter.

There are two cautionary factors to consider in buying heat pumps. One is that they don't make sense unless you need both a heater and an air conditioner. Buying a heat pump for just one of the two functions would not be economical. The second factor is that heat pumps work best in *moderate* winter climates,

and get less efficient at very low temperatures. At 15°
and below, some heat pumps automatically switch to
an electric-resistance heating device. But there are
heat pumps on the market that can keep working at
17° F. and are still twice as efficient as regular electric heaters, even at that low temperature. If you live
in a cold part of the country, you can check with local
appliance stores about the efficiency claims of the various available heat pumps.

With fuel oils getting scarce in some parts of the
country, heat pumps offer the benefits of electric reliability without the costly penalties of electric inefficiency.

Nevertheless, if you want to get completely away
from electricity—and you don't trust fuel oil or natural gas—then one of the very-improved wood-burning
stoves might be just what you need.

Wood Stoves

Most homes used to be heated with wood. Wood
went out of style, generally, when the fuel companies
went underground for their sources. People gladly laid
down their hatchets, wedges, and log splitters for a bin
of coal or a tank of oil.

Wood is making a comeback because of the high
prices of electric and oil heating, and also because
wood provides a kind of energy independence. You
don't need a drilling rig to discover wood; you can
find your own around forests, building sites, or land
developments. Even if you have to buy wood, the
prices are now competitive with gas and oil; and if
you harvest your own, you can reduce your heating
bill to almost nothing. If you live in an area with an
abundant supply, wood will always be a reliable
source of fuel. There won't be an Arab wood embargo.

Wood isn't gaining in popularity simply because
other fuels have problems. Three developments have
helped increase the efficiency and reduce the agony
of both collecting and burning wood. Those developments are: insulation, the chain saw, and advanced
stove design.

The Hidden Dividend

It used to take twenty cords of wood to heat a large, uninsulated home; and twenty cords made for a lot of sawing and splitting, not to mention huffing and puffing. A similar, modern, well-insulated home requires only five cords per winter.

The tremendous burden of cutting trees and sawing logs by hand has been lightened by the chain saw. Other devices, like the Roto-Wedge, even make log splitting into an easy, Sunday-afternoon pastime. The Roto-Wedge is a long, circular auger that attaches to the rear-wheel lugs of a car or truck. You jack up the back end of the car, spin the wheels, and the Roto-Wedge screws into the log, splitting it apart. The Roto-Wedge can be purchased from Perley C. Bell, Sales & Service, Grafton, Vermont 05146—who reports that Roto-Wedge can split 480 logs an hour.

Traditional fireplaces were as inefficient as the uninsulated houses they heated. The key to efficient wood burning was the invention of the airtight stove. It might not be as much fun to watch as a fireplace, but it keeps you a lot warmer—and on much less fuel. You can even control the rate of burning, by adjusting the air flow on these stoves. There are many good ones on the market.

The cutaway shown in Plate 4 shows the interior of one such stove, the Defiant Parlor Stove, manufactured by the Vermont Castings Company, Randolph, Vermont. The stove serves a dual purpose: you can cook on it, and can heat with it at the same time. In addition, with the doors open, it serves as a fireplace. Such a stove would be a perfect item for the survival room described in Section 9.

As you can see in the diagram, this stove is much more complex than a fireplace. It has various adjustments and controls. When the damper is open, as it must be when you open the doors to start the fire or to add wood, then the flames and smoke rises vertically up the chimney. This is exactly what happens continually in a regular, open fireplace. This traditional burning process is a waste of fuel. It is the equivalent to putting a single flame of natural gas in your fireplace, and using that as a heater. You would get the peripheral benefit of some of the heat given off by the flame, but the main thrust of it would be completely lost up the chimney. When the damper is turned down, after the loading doors have been closed, the Defiant Stove becomes a high-efficiency heat-producing chamber. The air comes in through an opening controlled by a thermostat, which can regulate the intensity of the fire. The air is preheated at the bottom of the chamber, which causes the wood at the bottom of the pile to burn very fast. Burning the wood on the bottom helps dry out the wood stacked on the top of the pile. As the bottom logs disintegrate, the logs above them fall, to take their place. When burning slows down, the thermostat can open the port to bring in more air and heat things up again. With this controlled burning process, the stove can run for up to eighteen hours on one load of wood.

A good wood-burning stove, such as the Defiant, does even more than that. It also *recovers* most of the hot gases released by the burning wood—gases which would normally go up the chimney—and burns them to make even more heat. In the Defiant Stove, these exhaust gases are ignited along a flame path that winds for sixty inches around the combustion chamber. Preheated secondary air is introduced to complete the burning process. At the same time as the hot gases are being turned into heat, the stove is burning off the creosote and other substances that would otherwise gum up the inside of the chimney.

Wood-burning stoves come in a variety of shapes, sizes, and designs. At the end of this section, we include a list of some 25 manufacturers of wood-burning products, along with the type of equipment they make. Here are a few of the more noteworthy or unusual items on the list:

The S.E.V.C.A. Stove is made from recycled propane tanks. It is a high-efficiency wood-burner sold at a very low price, and it was designed as a volunteer action project for low-income families. It has many of the same design features—such as secondary burning of exhaust gases—that are found in more expensive stoves. You can cook on it, or heat with it; and a water coil inside the upper chamber can even be used to provide hot water. Tests reported in *Country Journal* magazine indicate that the stove is capable of producing at least 50,000 BTUs per hour.

The Add-A-Furnace (Plate 5) can be placed alongside an oil or gas furnace, where its stovepipe connects into the existing chimney. When the wood-burner is fired up, a blower forces heat into the hot-air plenum of the existing furnace, then through the hot-air ducts into the house. It can be used as the primary heat source, or as a backup in case of a fuel shortage.

The Logwood wood-burner can automatically shift from wood to oil. The oil burner is protected from the wood fire by a grate that can be opened and closed. The grate is controlled by a thermostat, so when more heat is needed, the grate opens and the oil burner is automatically ignited. The manufacturer claims that this furnace offers very high efficiency. An independent testing laboratory rated its thermal efficiency at 84.2 percent; most conventional warm-air furnaces have an efficiency rating of about 75 percent.

The attractive Monarch combination kitchen range, with white-enamel finish, is a fine example of the modern wood-burning ranges now available that would fit into any contemporary home. Models utilize either wood or coal for both cooking and heating, in the winter; in summer, the gas or electric elements are used, which reduces the load on the air-conditioning system.

Kristia Associates, of Portland, Maine, imports the Jötul wood-burning stove from Norway. They asked us to point out the importance of regular chimney cleaning. They sell a booklet called, "The Chimney Brush," a homemaker's guide to the art of sweeping chimneys. The book is packed with useful information and is written in a humorous vein. You can get a copy by sending $1 to Kristia Associates, P.O. Box 1118, Portland, Maine 04104.

Stoves and Ovens

If it is worth the effort to eliminate the pilot lights you already have, it's certainly worth it to buy a new stove that has no pilot. Several manufacturers make these stoves. If you prefer to cook with gas, find a stove with electronic ignition.

If you can't decide between gas and electric for cooking, check the fuel prices in your area. Natural gas, as we have shown, has been by far the cheapest way to cook. (Studies are inconclusive, however, on the relative economics of electric cooking versus propane, or LP, gas.)

Once you have decided on gas or electric, the next problem is which brand to choose. No comparative-energy ratings are available for ranges and stoves, as there are for refrigerators. Ovens use more or less energy, depending on how much insulation is stuffed into their walls and doors. It is difficult to measure the insulation in an oven, but you might try asking your appliance dealer which ovens are the best insulated.

Self-cleaning ovens require more insulation, because of the high temperatures generated during the cleaning process; the cleaning itself costs extra energy money. Between the electricity lost in cleaning and the electricity saved during baking (because of these ovens' better insulation), you probably come out about even. If you have a self-cleaning oven and use the cleaning apparatus only infrequently, you may in fact get a relative savings from the extra insulation. By doing the cleaning immediately after the oven has been used, when it is still warm, you make an additional saving.

See Plate 6.

Manufacturers of Wood-Burning Stoves (for Heating and/or Cooking) and Related Products

Manufacturer	Products
Atlanta Stove Works, Inc. P.O. Box 5254, Station E Atlanta, Ga. 30307	Homesteader wood-burning heater; automatic wood-burning heater; wood-burning cookstoves
Bellway Mfg. Grafton, Vt. 05146	Wood-burning furnaces; Roto-Wedge

Manufacturer	Products
Brown Stove Works, Inc. P.O. Box 490 Cleveland, Tenn. 37311	Automatic wood-burning circulator
C & D Distributors, Inc. P.O. Box 766 Old Saybrook, Conn. 06475	Fireplace stoves: Jiffy Woodsplitter; Jiffy Chimney Cleaner
Fisher Stoves Highway 15, P.O. Box 481 Watkinsville, Cal. 30677	Wood-burning stoves
Hydroheat P.O. Box 382 Ridgway, Penn. 15853	Heat-saving fireplace; grate for hot-water systems
Hydraform Products Corp. Box 2409 Rochester, N.H. 03867	Eagle Stove with refractory firebox
Kristia Associates, Importers 343 Forest Ave., P.O. Box 1118 Portland, Me. 04104	Complete line of Jötul Stoves, imported from Norway
Lassy Tools, Inc. Plainville, Conn. 06062	Heat-Catcher forced-air fireplace grate
Malleable Iron Range Co. Beaver Dam, Wis. 53916	Monarch wood-burning Add-A-Furnace; kitchen heaters; kitchen ranges; Franklin Fireplaces
Marathon Heater Co., Inc. P.O. Box 365, R.D. 2 Marathon, N.Y. 13063	Logwood Wood-Burning Furnace and Boiler (switches from wood to oil automatically)
Martin Industries (King Division) P.O. Box 128 Florence, Ala. 35630	Wood-burning stoves
Mohawk Industries, Inc. 173 Howland Ave. Adams, Mass. 01220	Temp-Wood wood-burning stoves
New Hampshire Wood Stove, Inc. P.O. Box 310, Fairgrounds Rd. Plymouth, N.H. 03264	Wood-burning stoves
Portland Stove Foundry, Inc. 57 Kennebec St, P.O. Box 1156 Portland, Me. 04104	Wood-burning parlor stoves and kitchen ranges; Franklin Stoves

Manufacturer	Products
Riteway Manufacturing Co. Division of Sarco Corp. P.O. Box 6 Harrisonburg, Va. 22801	Wood- and coal-burning stoves, furnaces, and boilers
Rocky Mountain Stove Co., Inc. 5231 Chinden Blvd. Boise, Ida. 83704	Timberline stoves and fireplaces
Scandinavian Stoves, Inc. Box 72, Route 12-A Alstead, N.H. 03602	Complete line of Lange Stoves, imported from Denmark
S.E.V.C.A. Stove Works P.O. Box 396 Bellows Falls, Vt. 05101	S.E.V.C.A. low-cost, high-efficiency space heater and stove
Suburban Manufacturing Co. P.O. Box 399 Dayton, Tenn. 37321	Woodmaster Space Heaters
Superior Fireplace Co. 4325 Artesia Ave. Fullerton, Cal. 92633	Heatform energy-efficient fireplaces and accessories
The Merry Music Box 20 McKown St. Boothbay Harbor, Me. 04538	Complete line of Styria cooking ranges and heaters, imported from Austria
Thermo-Control Wood Stoves Cobleskill, N.Y. 12043	Thermo-Control wood stoves, water heaters, Franklin Fireplaces, and hot-air heaters
Vermont Woodstove Co. P.O. Box 1016 Bennington, Vt.	Wood-burning space heaters
Washington Stove Works P.O. Box 687 Everett, Wash. 98206	Wood-burning stoves, parlor stoves, cooking ranges, and heaters

Hot-Water Heaters

Hot-water heaters traditionally lack good insulation, but some of the newer ones are made with a better blanket. If you have to buy a new hot-water heater, look for the one with the thickest and best insulation that is available. There isn't much difference in the heating elements of electric water heaters, so the energy-saving possibilities all depend on the insulation of the tank.

The most economical gas heaters—from an energy standpoint—are the ones with electronic ignition instead of a pilot light.

If you are installing a new water heater, where you put it can be just as important as what kind you buy. The long pipe runs described in Section 2 can be easily eliminated if you locate your heater near the kitchen and bathrooms, where most of the hot water is used. That way, very little heat will be lost to the pipes, and extra pipe insulation will not be a necessity.

Washing Machines

Find out how much water is needed for a complete cycle, and buy a machine that uses less water. Some machines even have flexible water levels. Others have fast spin speeds that reduce drying time. Suds savers and mini-baskets are also good energy-saving features.

Dryers

Electric dryers are all about the same, in terms of energy efficiency. Machines that have Fluff or Air-Dry settings can save a little electricity, but not enough to worry about. If you are going to buy a gas dryer, however, look for one with electronic ignition, so you won't have to support another eternal flame.

Dishwashers

There are some energy-saving models on the market that use less water and have shorter cycles. Make sure the one you buy has a switch to turn off the automatic-dry cycle; air-drying is effective and it doesn't cost anything. The benefits you can derive from an energy-saving dishwasher are very small compared to those realized with an energy-saving refrigerator or air conditioner, however.

Televisions

Energy-saving television sets are now on the market that use less than 100 watts of power each hour. Before you buy such a set, compare its price with that of a regular color set. Unless you watch a tremendous amount of television, the savings on electricity may not be enough to cover the additional cost of the energy-saving color set.

Waterless Toilets

If you want to save even more water than you are now saving, or if you live in an area that does not allow any new sewer hookups, there are a couple of toilets on the market that use no water at all. They are expensive.

One, called Clivus Multrum, turns human and kitchen waste into compost that can be used in the garden. It processes the waste material entirely automatically, reducing the original volume by 90 to 95 percent. It does this without any machinery or moving parts. The complete system costs $1,685, and it would be ideal in an area where septic tanks can't be used. The system has been in operation for more than twenty years in Sweden, and is distributed in this country by Clivus Multrum U.S.A., Inc., 14-A Eliot St., Cambridge, Mass. 02138.

A similar system, called Humus-Toilet, is distributed by Future ECO-Systems Ltd., 680 Denison St., Markham, Ontario, Canada L3R1C1.

Further Information

One way to get the most energy-efficient appliances is to buy the same ones that the government does. The General Services Administration uses life-cycle costing in its purchases of appliances for government installations, and you can get the complete list from: Consumer News Room 3300, 300 Independence Ave. S.W., Washington, D.C. 20201.

The list contains only the appliance models of companies that have bid on government contracts, so some efficient appliances may not be included on it.

SECTION 6

Tightening Up the House: An Extra Blanket Doesn't Always Pay

Introduction

So far, we have presented the adjustments in hardware that make appliances use less fuel; the adjustments in behavior that make people use fewer appliances. If you have followed a plan, you have done some of the no-cost things, considered some of the low-cost things, and, hopefully, you have saved your first 25 percent. The next 25 percent gets harder. From this point on, energy saving enters more expensive territory.

Insulation, solar water heaters, solar space heaters, and heat-recovery units are some of the major items we cover in this section and the next two. The installation of these and other such products has been given the space-age name of "retrofit"—which sounds exciting, as if you are preparing your home for takeoff. The prices for retrofit items have also taken off recently—which is reason enough for caution. With the exception of caulking and weather stripping, which you can easily do yourself, the other energy-saving modifications cost in the hundreds of dollars, and solar heating can run into the thousands.

Some of these items, expensive as they are, turn out to be terrific investments, nevertheless. Others are not so good. We will try to help you make the best choices for your particular situation. The choices have been made more difficult by the recent bombardment of the American market by energy-saving products, and by the overly optimistic claims of some promoters of solar heaters and insulation. You can often add up the projected savings of insulation, weather stripping, storm windows, and solar heaters—and those "savings" exceed the amount you are actually paying for fuel! So something must be wrong. You can't save $400 a year on attic insulation, $75 on storm windows, $150 on basement insulation, and $250 on a solar space heater if your fuel bill is only $500.

This excessive optimism has not been due to blatant misrepresentation. The projected savings you hear about are so dependent on the special conditions in each separate house or apartment that there is hardly a way to generalize. And the savings estimates for various retrofit items usually leave out an important factor: the installation of one energy-saving device can reduce the gains made by another. If you decide, for instance, to invest $500 in a wood stove to take advantage of a cheaper source of fuel, then the return you get from extra insulation will be reduced, because that return is determined by the cost of fuel. On the other hand, if you insulate first, you may be able to use a smaller and less expensive wood-burning stove.

Such are the complexities that make retrofit a tricky business. It is important that you know your fuel costs and your electric costs, for both heating and cooling. They are the key to whether any of the rest of this makes economic sense. It is also important that you consider all the items presented in the next three sections before you decide where to put your money. We have discovered that some completely unpublicized products, such as the heat-recovery unit, can often offer a better return, and for a smaller investment, than glamour items like the solar water heater.

Insulation

If your house is built like most houses, it, itself, wastes more fuel or electricity than all the oven peekers, refrigerator starters, and thermostat jigglers do who inhabit it. Corrective actions like insulating the attic or putting up storm windows have the advantage of not requiring daily vigilance or nagging by the energy manager. The benefits do not depend on human behavior, and in spite of all the energy gluttony that may continue to take place inside the house, the insulation will still keep working. And insulation has a double payback: it saves heating money in the winter and cooling money in the summer.

Even with all these advantages, insulation is not always a great investment. The general advice on insulation has been to pile on as much as you can afford. This may be helpful from a national energy standpoint—the more insulation you have, the less fuel you use!—but not necessarily from a personal economic standpoint. There comes a point of diminishing returns, a situation similar to what happens when you pile blankets on a bed. The first blanket is crucial, the next one makes some slight difference; but after that, who can tell? When you are building a new house, it makes sense to use as much insulation as possible, for reasons which we will discuss in Section 13. But if you are modifying what you already have, caution is advised. Some other action may turn out to be more profitable than the extra blanket.

The problem with insulation is that nobody can tell you exactly how much more you need. General estimates are to be found in one book or another, but such estimates are meaningless. There is just too much variation in fuel costs, in furnaces, in types of houses, and in weather conditions in order to make a useful across-the-board projection of savings from this or that amount of insulation. The State of Massachusetts directed an ambitious statewide pilot energy project to encourage people to insulate their homes, but the people there had two great advantages in deciding how much insulation to purchase. The first was the infrared scanner, which takes thermal pictures of houses and shows the places where heat is being lost. (A scanner is much too expensive for any individual to buy, but if your town could rent one—the scanner is discussed in Section 12—it might help with insulation decisions.) The second was a computer analysis, offered free by the federal government; people could fill in a questionnaire, with the details about their own house and its furnace, and the government would sent back a complete economic analysis of the costs and savings of insulation, storm windows or weather stripping.

Unfortunately, this service is not now available on a nationwide basis, but we think we have found a useful substitute. It is a booklet called "In the Bank or Up the Chimney," published in 1975 by the Department of Housing and Urban Development (HUD). You can get a copy by sending $1.70 to the Superintendent of Documents, U.S. Government Printing Office, Washington, D.C. 20402; the catalogue number for the booklet is 023-000-00297-3. (For some curious reason, the identical publication is available at some bookstores under the title, "How to Keep Your House Cool in Summer and Warm in Winter." In this form, the booklet sells for $2.95!)

The HUD booklet does more than just estimate the savings you might get from insulation or weatherstripping. It provides an elaborate worksheet, so that you can make your own calculations. The worksheet includes such factors as the amount of money you pay for fuel or electricity, the kind of house you have, and the area of the country you inhabit. It takes a couple of hours to run through the whole process, but people who take the time to do it will end up with the most accurate retrofit analysis that they could get anywhere. With the booklet, you can compute the costs and benefits of storm windows, basement insulation, wall insulation, and turning down your thermostat. The booklet also tells you where to buy the various items, what to buy, and how to do the installation yourself.

The booklet can get a little troublesome for people who don't like to fill out forms, or who feel timid around decimal points. We have, therefore, in this Section (6) devised our own simplified procedure for evaluating insulation and weather stripping, etc., drawing on material from the booklet and from a variety of other sources. The payback figures for each retrofit item will vary from house to house, as you will see, but we have found that the *relative* merits of each action stay about the same; that is, fairly definite conclu-

Tightening Up the House

sions can be drawn as to which modification brings the greatest return, which is second, and so on.

To illustrate our procedure, we will take a fairly typical house in the Philadelphia area. We have chosen Philadelphia because a great number of people live in its same climate belt, away from the extremes of heat and cold. The house has 1,200 square feet of living area and was built in 1958. It has benefited from only a minumum amount of maintenance, and has no insulation. There is an unfinished basement with an oil-burning warm-air furnace; a central air conditioner has been added. The owners have to struggle to heat this house. They report that even when they set the thermostat at a chilly 64°, they burn at least $860 worth of fuel in a heating season.

What should they start by doing?

Insulating the Attic

For every house, and for every area of the nation, the heating and cooling factors will be different. But the conclusion is always the same: The first thing to do is to insulate the attic, if it has no insulation already. More heat is lost there than anyplace else in the house, since the hot air from the furnace rises to the top floor and is cooled by the frigid top-floor ceiling (attic floor). The infrared scanner would show the uninsulated roof cavity of the Philadelphia house to be one big "hot spot." In some places poor insulation can be detected without the scanner: when the snow melts quickly on the roof! The heat that is supposed to be keeping the Philadelphia family warm is keeping the roof warm, instead.

If they do it themselves, they can insulate the attic of this Philadelphia house for $259.00. (Insulation costs will continue to go up, so check in your area for current prices.) The various calculations in the HUD booklet, adjusted for inflation, tell us that the payback will be $430 a year. The insulation itself is thereby paid for in less than one year, and the family will begin to show a profit already. If they hire a contractor to install the insulation, they should still break even after the first year. Over the next thirteen years, the dividends from this retrofit will add up to $5,331, the "In the Bank..." analysis tells us.

The dividends seem too good to be true, but insulation in an attic can be that effective. Attic insulation, remember, will also save on fuel bills in the summer. In fact, in warm climates, the insulation will cut more off the air-conditioning bill than off the heating bill. A good layer of ceiling insulation will keep the summer heat from getting into the house, and, with an attic fan to release some of the heat to the outside air, the cooling bills will be drastically reduced. We have included the cooling factor in our projected savings for the Philadelphia house.

In our example, we have gone from no attic insulation to a level of R-22. Insulation is sold by R-value, which is simply the amount of insulating power that a particular substance has. The higher the R, the better the insulation. However, the research done by Dr. Jay McGrew in Colorado has underscored the fact that insulation actually installed in houses might not perform as well as its R-value would imply. The more insulation you add, McGrew says, the greater the discrepancy between the theoretical insulating power and the real results. His conclusions are based on factors like moisture levels in your home, factors that we cannot define. Our advice is to check with people in your area who have installed insulation, to see if they got the expected fuel savings for the R-level of insulation they bought. If you ask around, you will get a pretty good idea of whether insulation is living up to its advertised effectiveness. You can't just assume that R-values will be a valid measure. On the other hand, these values are the only way we know of comparing levels of insulation, so we will rely on them in this analysis.

Ceiling insulation is sold at building-supply stores, and comes in batts, rolls, or bags of loose or shredded material. Try to get the most Rs for your money. Also try to get insulation that has been proven not to be flammable. (Plastic insulations have been criticized in this regard.) The rolled fiberglass ceiling insulation with an R value of 22, which was used in the Philadelphia house, is six inches thick and has an aluminum vapor barrier on one side. It is easy to install. You just lay it between the joists in the attic, with the aluminum backing down. (Remember to caulk around the lighting fixtures and the chimney stack before applying the insulation.)

The loose-type insulation can be spread around the attic, but it is messy. Contractors can blow loose insulation between the joists without too much trouble; so if you want that type of insulation, you might consider letting a contractor do it.

For finished or partly finished attics, where a floor already exists, installing insulation is a more difficult proposition. You don't want to rip up the floor, so you will have to get a contractor to blow the insulation into the cavity between the floor and the ceiling below. If you have a finished attic floor but an unfinished roof, you can tack insulation between the roof joists, but such a procedure will not be as effective as insulating the attic under its floorboards.

How much insulation should you put in your attic? The HUD booklet recommends that if you start with no insulation at all, you should put in R-22, or six inches of fiberglass. If you have electric-resistance heating in your house, or have oil heat and live in a very cold climate, then they recommend even higher levels, such as R-30 or R-38. The McGrew study in Colorado would lead us, however, to question whether this extra amount beyond six inches is really worth the money. Again, it might be; but you should check with other people in your area before buying.

To go from no insulation to R-30 takes about thirteen inches of fiberglass, or eight inches of cellulose fiber. And of course, for an add-on project, you must use insulation that doesn't have a vapor barrier. Your building-supply dealer can help you figure out how much to buy to get the R-value you require. He can also tell you the pros and cons of the various materials that are available.

If there is no insulation in your attic, buying a few rolls of insulating material will bring the kind of return that gamblers dream about. In our opinion, this is an investment that cannot help but succeed—regardless of where you live and what kind of house you have. Our optimism, however, cannot be automatically applied to *adding* attic insulation, if you already have some. The virtues of insulation are so well publicized that people frequently make the mistake of thinking they will get the gambler's returns for *extra* insulation. They probably won't. In a Princeton retrofit project, attic insulation was raised from a level of R-11 to R-30. The researchers estimate that the payback will occur only in eighty-four months, or seven years.

If we assume that our Philadelphia house already has four inches of insulation in the attic floorboards, then the savings factor is reduced. Instead of a $430-a-year payback, they will save only $43 a year, or one-tenth the amount. The cost of installing the insulation is reduced by about half, of course, and the break-even point is reached in three years. That still isn't bad, but the $43-a-year benefit has to be carefully weighed against the potential benefits from other retrofit projects. We can't tell you exactly what you will get back when you *add* attic insulation, but we can suggest that it will not be spectacular. In any case, the HUD booklet suggests the following optimum insulation levels:

Table 9: Optimum Attic-Insulation Levels

Thickness of existing insulation	How much to add	How much to add if you have electric heat; or if you have oil heat and live in a very cold climate	How much to add if you have electric heat and live in a very cold climate
0"	R-22	R-30	R-38
0"–2"	R-11	R-22	R-30
2"–4"	R-11	R-19	R-22
4"–6"	none	R-11	R-19

Insulating the Basement

An uninsulated basement or crawl space can be a major souce of heat loss. In the Princeton townhouse retrofit study, basement losses were substantial. The Princeton people estimated that 25 percent of the total furnace heat ended up in the basement of the townhouses they monitored! If your furnace is located in a cold basement, and if the first floor above the basement is chilly even when the heater is turned up, then you have a profitable retrofit opportunity.

There are several ways you can go about insulating a basement. The easiest and most profitable, according to the HUD booklet, is to insulate the floor above the basement. We could do the floor in our sample Philadelphia house for a cost of $132, and the payback would be $239 a year, for a thirteen-year total of $2,975. In this example, the figures give insulating the basement a higher priority than *adding* insulation to an attic that already has some.

It is just about as easy to insulate a floor above the basement as it is to do the attic. You lay the rolls of insulation in between the floor joists, from below, with mesh or chicken wire stapled to the bottom of the joists, to hold it in place. It takes a little time, but the results will pay you a terrific hourly rate.

Once you have decided to insulate your basement, don't forget about the furnace ducts. It doesn't make much sense to leave uninsulated ducts in a cold basement, especially when the heat that escapes from the ducts can't get through the floor insulation to help warm the house. It won't cost you much to put insulation around the furnace ducts, and you will lose a lot of heat if you don't.

If you have a crawl space, it's also a good idea to insulate that. One way to do this is to tack insulation along the sides of the crawl space and put a vinyl plastic barrier on the ground underneath. A local builder can advise you on the best approach for your situation. We figure that if our Philadelphia house had a crawl space instead of a basement, we could insulate it for $77, and the return would be $59 a year, for a thirteen-year profit of $690.

Some people like to use their basement as a workroom, machine shop, or playroom for the children. If you plan to make your basement into a functional room, you may prefer to insulate the basement walls and caulk and weather-strip the windows, instead of insulating the basement ceiling. When a heater or furnace is located in a fully insulated basement, any heat that is lost to the basement will be put to a dual purpose: It will make the basement comfortable, and it will also warm up the first floor. Since feet are one of the principal body thermometers, a toasty floor can help keep the thermostat down.

Insulating the walls of the basement is a much bigger job than insulating the floor, however, and the returns are not as promising. In our sample house, we could insulate the walls for a cost of $422 and get back $144 a year, or $1,450 on the thirteen-year life-cycle calculation. That's still a good investment, but not as good as when we insulated the floor over the basement, and got back $2,975 over thirteen years for an investment of $132. You will have to decide in your own case whether the use of the basement is worth the extra money.

If the basement is too hot after you have insulated the walls, then you should insulate your furnace ducts. In the New Jersey houses involved in the Princeton study, so much heat got into the basement that there was enough left over to keep the first floor warm even *after* the ducts were insulated. After the ducts were insulated, more heat was transported to the other areas of the house.

Insulating the Exterior Walls

We could save $150 a year by insulating the walls of our sample house, but it is very hard to insulate walls that have already been sealed. The job is an expensive one, requiring the services of a contractor, and it would cost us an estimated $614 to do it. The contractor has to cut holes in the outside wall between each pair of studs, and blow in mineral fiber, cellulose fiber, or urea-formaldehyde foam insulation. There is no way to be sure that a satisfactory job has been done. The insulation can hang up on electrical wiring or other obstacles in the walls, and leave a large void below. It has been estimated that up to 30 percent of walls filled by this method remain uninsulated. This reduces the fuel savings by the same amount.

It would be possible to check for voids with the use of an infrared scanner, but because of the high cost of this equipment it would hardly be practical for an individual homeowner. Furthermore, some types of blown-in insulation tend to compress, leaving a void at the top of the wall, and increasing the heat losses even further.

Plugging the Leaks

Even in a house with good insulation, a lot of furnace heat can get out through cracks around the windows, or through small holes in the walls, or from spaces between the floor and the baseboard. These air leaks are called exfiltration if heated air is escaping, and infiltration if cold air is coming in.

It is hard for many people to visualize how a few little cracks and openings can have much effect on a fuel bill. But they can. As easily as a tiny hole in a shingle can cause major water damage through a roof, a crack around a window frame can provide an exit for a major mob of heated air particles. In the average house without caulking, the inside air may in this way be completely flushed out and replaced every fifteen minutes. In high winds, the air changes can happen even more frequently. Every time one of these air changes occurs, all of the new air must be heated from the outside temperature up to whatever level of coziness you have set on your thermostat. That's, ordinarily, four new batches of air to heat each hour. If it is *very* cold outside, the cost of this continuing process can be tremendous.

Most leaks can be stopped with caulking and weather stripping.

Caulking

Not only is caulking an easy job for an amateur, but in the process of investigating leaks you will get a better idea of the thermal characteristics of your house. You will learn to spot the cold places and the drafty places that may require insulation or other treatment later. Caulking is also relatively inexpensive. That is why we rank it second to insulation, even though the returns from other projects might be greater.

It is usually assumed that most of the air leaks in a house happen around the window and door frames, but new information suggests that other places can be just as porous. The Texas Power and Light Company did an air-leak study, and reached the surprising conclusion that, in a home of 1,728 square feet, 25 percent of the air escaped around the soleplate (the space between the floors and walls); 20 percent leaked through the wall outlets; 13 percent leaked around windows; another 13 percent through the duct system; 6 percent through the hood vent,; 6 percent through the fireplaces; 4 percent around exterior doors; 3 percent through the dryer vent; 1 percent through the bathroom vent; and 5 percent through recessed spotlights. These estimates are sure to vary from house to house, but the point to remember is that doors and windows are not the only sensible candidates for a caulking job.

In his Denver research, McGrew made similar findings. He identified five major holes that cause the majority of the infiltration. The holes were: the kitchen exhaust, the bathroom exhaust, the hole around the dryer vent, the hot-water intake, and the furnace flue. While weather stripping around doors stopped about one-tenth of an air change in the house per hour, the five holes were responsible for five air changes per hour in the Colorado houses that he studied. If you are tightening up the house, give plenty of attention to these holes, and to any other large surfaces where air might be leaking in. You can tell the relative contribution that any crack or hole makes toward cold-air infiltration by calculating its size. A tiny crack along the length of the door will not equal the heat-loss potentiality of an exhaust vent.

You can buy caulk, at varying prices and quality, from the local hardware store. It is applied with an inexpensive caulking gun, much as decorative icing is spread on a cake, in a long and continuous stream of compound called a "bead." The better grades of caulk can last as long as ten years; the cheaper grades tend to harden and crack, and require frequent re-caulking.

Mechanix Illustrated published an article in September 1976 called "Consumer's Guide to Caulking," which gives a complete analysis of caulking compounds and a detailed description of how they are applied. In general terms, oil-based caulks are cheap, but give poor performance. Vinyl compounds last from two to five years, and can't be used on concrete. Butyl caulks last about the same length of time, but tend to shrink; and they don't produce a particularly good "bead." The best caulking compounds are silicone and acrylic latex. A standard eleven-ounce tube of silicone or acrylic latex will cost between $2 and $4, but the caulk will stay in place for more than ten years. One tube will seal about twenty to twenty-five linear feet with a 1/4-inch bead. Some caulks are even paintable. General Electric makes an all-weather paintable silicone.

The most effective caulking is done outside the house. Some gaps between the floor and the walls can be quite wide, so you must force in oakum or other filler before you add the caulking compound. You should caulk the seam where the walls meet the foundation all the way around the house. While you are outside, put another bead of caulk along the seam where the chimney and the walls come together. Look for gaps around the siding of the building. Fill spaces around storm windows, where the metal frames join the house. Caulk the entrance holes for telephone wires or plumbing fixtures. Check around the edges of windows and the frames of doors, to see if those places have already been caulked. If there is caulking, but it is cracked, damaged, or brittle, it should be replaced. If there is no caulking, you should definitely use some, especially where there are obvious gaps between frames and walls. Caulking can be applied around the inner frames of wooden windows, as well as around the seams where the window casings join the outside walls.

For extra protection, you can also caulk around window and door frames *inside* the house, especially on the cold, windward side. And while you are at it, check for caulking in the attic area—before you add insulation. Seal around the chimney shaft as it passes through the attic. (A Princeton study found that a lot of furnace air escapes around that tiny space between chimney shaft and attic floor.) It is also useful to caulk around the recessed light fixtures, from the attic side. These fixtures are notorious air leakers.

There are many other places to caulk, and the most useful advice we can give is to use your own common sense. Any place where the outside light shines through, or where you can feel air drafts with your hands, or where you just believe leaks might occur, are good candidates for the caulking gun. Caulking is not so expensive that you need to be stingy with it.

Weather Stripping

After you finish caulking, you can go right into filling and weather stripping.

It is particularly important to fill up those gaps where the electrical outlets are cut into the walls. Turn off the power for the area where you are working, remove the wall-socket plates, and stuff oakum or small pieces of insulation all around the sockets. The Texas study showed that 20 percent of the air that leaked out of a test house came through those sockets. Fill any other holes you have discovered.

Weather stripping seals up the surfaces where a door or a window opens and closes. There are many types of weather stripping. The less durable, cheaper kind is adhesive-backed foam. It works okay, but not on moving surfaces—where it wears out quickly. Rolled vinyl with aluminum channel backing and spring-metal weather stripping are both very durable. Some types are harder to install than others. It is easy to install rolled-vinyl weather stripping, but it can be seen around the window. Check with your hardware store to find out what is available, how difficult it is to put in place, and how much it costs.

There should be weather stripping around all outside doors and windows, especially on the windy sides of a house. The Princeton study revealed another important place to weather strip: around the metal tracks that hold sliding glass doors and windows. A better sealer around these openings, the study says, can reduce infiltration as much as 50 percent. They suggest strips of closed-cell foam plastic, which squeeze around the door frames and keep them tightly sealed. This material is estimated to cost $25, and will provide a payback in twenty months.

In the Philadelphia house we described at the beginning of this section, an investment of $74 in caulking materials will provide a savings of $109 a year. Over a thirteen-year period, the savings will be $1,343. Not bad for a few tubes of gooey material. The "In the Bank . . ." booklet estimates that an investment of $75 to $105 in caulking and weather stripping will return $30 to $75 a year in reduced heating costs and $20 to $50 a year in smaller air-conditioning bills. The longest possible payback is two years, and in most cases you will break even before that.

Tightening the house offers an additional benefit. The outside air that seeps in during the wintertime is very dry. People feel chilly in dry air. By plugging the leaks, you keep more of the moist air inside, where it helps you remain warm.

Installing Storm Windows

Storm windows don't look like a very good investment in the HUD "In the Bank . . ." analysis. The permanent ones would cost us $324 to install on all the windows in our sample Philadelphia house, and the economic benefit would only be $49 a year, or $313 after thirteen years. If you are unsure whether an investment in storm windows would be worth the money, or if you don't own the house or apartment you are living in, we have a suggestion: Put up plastic storms, try them out for a year, and see if they lower the fuel bill. Plastic storms can be taped over the windows for less than 50 cents per window, and according to "In the Bank . . ." can save $20 a year in fuel bills. In the sample house, it would cost us $5.40 to install plastic storms, and we would get back $49 in the first year.

If you have good results from the plastic, you may decide to put up the permanent kind of storm windows. If you rent an apartment, maybe you could convince the landlord that such an investment would be in his own self-interest.

Two types of storm windows are available: the kind that must be removed every year; and the permanent, sliding kind that can be left on the windows all year round. The take-down storms cost about $12 a window, and the permanent units are $36, but the fact that you never have to mess with the second type again may be worth the extra money.

Storm windows are somewhat useful in keeping cool air inside the house in summer. The permanent ones include a screen, and can be opened in the summer if you are not using your air conditioner, so they have year-round applicability.

Storm windows have an additional benefit that goes beyond the pocketbook. In houses with single-pane windows, a cold winter day makes the glass just about as cold as the outside temperature. Body heat is radiated to this cold surface, making it very uncomfortable to sit or stand anywhere near the window. In effect, this reduces the living area of the house by a significant amount during the winter. When you put on storm windows, you recapture this area, and at the same time greatly improve the livability of your house.

You can also cut heat losses through your windows by double-glazing, or installing thermal-pane glass. For existing housing, this process is more expensive than putting on storm windows. Thermal panes are also less

Tightening Up the House

effective than storm windows, because the storm windows cover the entire window frame outside, and provide a complete seal around the window. Even with the thermal-pane glass, air can leak in around your window frames.

Storm Doors

Storm doors are not discussed in the "In the Bank . . ." analysis, but they can save you a small amount of money. The problem with storm doors is that most of the heat is lost when people are entering or leaving the house, not when the door is closed. The airlock, or vestibule, is a better solution. The airlock door is closed before the door to the house is opened, so furnace heat is never released to the outside. Many people have constructed makeshift vestibules out of polyethylene film and lath—and they report good results. Plans for more eye-pleasing, removable airlocks were discussed in the September 1975 issue of *Popular Science* magazine.

A Plan of Action

Based on what we have learned from the "In the Bank . . ." analysis, our suggestion for a plan of action would be as follows:

(1) During the first year, insulate the attic (if it has no insulation), caulk and weather-strip, and put up plastic storm windows. That would cost about $338, and would save $588 for our Philadelphia house. If the attic already has some insulation, then insulate the crawl space or the basement instead.

(2) After the first year, calculate the amount you have actually saved. The dividends should go a long way toward paying for the secondary projects, such as adding some bulk to the attic insulation or putting up *permanent* storm windows. But before you undertake any of the marginal projects, consider the items described in Section 7. It may be that the purchase of a heat-recovery unit or a solar greenhouse, will make a better investment than the purchase of more insulation.

Whatever you choose to do, don't try to do everything at once. Pick out one or two items, live with them for a month or two, and evaluate the results. If you can pay part of the cost of one retrofit item with the savings from another, the economic burden of your energy conservation will be lightened.

SECTION 7

Working Up to Solar

Before you put your money into a solar water heater or space heater, you might want to consider two other retrofit items that have received significantly less attention.

The first is the solar greenhouse, a product of a branch of energy conservation called "passive technology." The idea behind passive technology is to let the sun or wind heat or cool your house in their own natural ways, without processing them through complicated and expensive devices like collectors and windmills. Passive technology is a return to the obvious. Every house with a window is a solar house and a wind-cooled house, in the sense that at least some sunlight and some breezes can get inside. The passive technologists want to maximize these natural effects, especially in new house design. But you can do the same thing in an existing building if you add on a solar greenhouse, described below.

The second item is the heat-recovery unit, one of the first inventions to successfully connect one appliance into another. One of the problems with the American home is that all the energy functions are individualized. The refrigerator and air conditioner *give off* heat, while the water heater *requires* heat; but there hasn't been a way to take heat from one and transfer it to another. The heat-recovery unit does just that; and just as the solar greenhouse may sometimes be a better alternative to the solar space heater, the heat-recovery unit may be a more economical purchase than a solar water heater.

We also will make passing mention of windmills. In our opinion, they are not practical energy-saving devices for the average home.

The Solar Greenhouse: A Simpler Collector

It is possible to avoid the whole complicated solar-collector system and get similar results with a solar greenhouse. A greenhouse traps the sun's heat just like a rooftop collector; in fact, the whole solar process frequently is called the "greenhouse effect." You can do without the more complex solar collector's pipes, pumps, and storage tanks by attaching a small greenhouse building right at the side of your home. During the day, the greenhouse collects and retains a tremendous amount of heat, which can be circulated through the house to augment the furnace air. At night, you just close the door that connects the living area to the greenhouse, and turn up the furnace if necessary.

A solar greenhouse cannot produce the high temperatures reached in a rooftop solar system, but it has some advantages. First of all, it is much cheaper. You can buy a greenhouse kit for about $500, or build your own for less than that. Second, it provides a place to grow vegetables and plants within arm's distance of the easy chair. You can even weed during television commercials. The indoor garden can add an esthetic dimension to a home, especially when tomatoes are ripening inside while the snow is piling up outside. The greenhouse will thereby cut down on grocery bills and will occupy the family gardeners. Third, the greenhouse will add moisture to the air. Moist air can make people feel comfortable at a lower temperature, so the greenhouse may indirectly influence you to lower the thermostat. Not only will it produce heat, it will help you get by on *less* heat.

The Garden Way Publishing Company sells plans for a do-it-yourself solar room that was built by an engineer and an architect for their own homes. The plans cost $9.95 and can be obtained from Garden Way, Dept. 130ZZ, Charlotte, Vermont 05445. Garden Way estimates the following potential fuel savings of their solar room in certain specified cities:

Boston	$228 per year
Burlington, Vt.	77 "
Chicago	130 "
Cincinnati	124 "
Denver	224 "
Minneapolis	97 "
New York	325 "
Philadelphia	207 "

Several companies also produce prefabricated-greenhouse kits that can be attached to an existing building. Lord & Burnham, at Irvington-on-Hudson, New York, sells various types and sizes of greenhouses that start at $550.

Redwood Domes, Box 666, Los Aptos, California, sells one for $450, and Sturdi-Bilt Co., 11304 SW Boones Ferry Road, Portland, Oregon, markets a prefab greenhouse for $460. Depending on local building codes and your own cultural predilections, a greenhouse may be the best way for you to get some free heat.

The Heat-Recovery Unit: What Can the Sun Do That an Air-Conditioner Can't?

Before you consider harnessing the sun for your hot water, you might want to investigate harnessing your air conditioner. The device that does it is called a heat-recovery unit. It is less cumbersome, less complicated, and less expensive than a solar water heater.

The heat-recovery unit is a small piece of equipment that is attached to a central air conditioner and connected to the hot-water system. As long as the air conditioner is being used, the unit provides all the free hot water anybody could need. In the process, it improves the efficiency of the air conditioner and reduces its operating costs by 10 percent. With a heat-recovery unit, you *can* end up with no electric bills for heating water, and lower electric bills for cooling the house.

Waste-heat recovery is based on a simple premise. We talk about how air conditioners "cool" the house, but what they actually do is remove heat from the inside and dump it outside. Meanwhile, perhaps not far from the air conditioner, we employ a water heater to produce heat, and at very high prices. Somebody had to figure out a way to take the waste heat from the air conditioner and use it to produce hot water. If you have a central air conditioner, and live in an area where it is in operation for five to six months of the year, a heat-recovery unit may be an investment far superior to a solar water heater. It will cut your hot-water bill to zero for about half the year—which will give the same savings as a solar heater that produces 50 percent of your hot water for the whole year. And, if your air conditioner happens to be a heat pump, the heat-recovery unit will work all year, elevating it from a good to a spectacular investment!

We first heard about heat-recovery systems from an article by Evan Powell in the October 1975 issue of *Popular Science*. He reported on a system designed and installed by Gene Chandler, the city electrical inspector for Decatur, Alabama, for his own home. Chandler has a heat pump, so the recovery unit works all year round. He built it from junk parts for about $200. Chandler tells us that the system has supplied 100 percent of his domestic hot water for the past two and one-half years, and has paid for itself long ago.

Through further investigation, we discovered that heat recovery for hot water is not a new idea. One of the first systems was installed in the Colonnade Restaurant in Tampa, Florida, in 1959. Since that time, a Florida utility, the Florida Power Corporation, has continued to develop the equipment. One of their spokesmen looks at it this way: "We don't like the idea of installing one piece of equipment for the purpose of removing heat, and at the same time installing other equipment to add heat." As a result of the utility's program, hundreds of heat-recovery units are being used in restaurants, supermarkets, dairy plants, and other commercial establishments. Engineers have been installing scaled-down versions of the big commercial units in their own homes. Factory-made units have become available for residential use only recently. As of now, a few thousand homes in central Florida are using heat-recovery units, but the rest of the country is still virtually unaware of the existence of this worthwhile device.

Heat-recovery units for homes have not been publicized, we think, because they can't compete with the glamorous and poetic image of solar heating. The process of capturing the sun and taking some advantages of it gives people a Promethean sense that they are beating the solar system. The process of capturing heat from a mundane air conditioner has no such mythological overtones. On the other hand, it may be more rewarding. The installation of a solar hot-water system can cost somewhere between $1,000 and $2,000, and will probably pay off in between six and twelve years. A heat-recovery unit can be installed in a new home for less than $300, and in an existing home for $500 to $600. Instead of huge, weighty collectors, the heat-recovery device is contained in a box that is 17 inches long, 11 inches wide and 6 inches high. The installation involves running two tubes from the heat-recovery unit to the hot-water tank and making two connections to the tubing that joins the air-conditioner compressor to the outdoor condenser coil—plus the electrical connection.

Plate 7 shows, diagramatically, how heat recovery works. A refrigerant solution moves from one side of the air conditioner to the other, changing back and forth from liquid to gaseous form. The liquid is released into the indoor coils, where it evaporates, removing heat from the air in the process; that is what cools the house. By the time the refrigerant leaves the indoor part of the air conditioner, it has been changed into a gas. The gas, which contains the heat extracted from the house, is compressed, and its temperature soars to over 200°. The heat-recovery unit uses this superheated gas to raise the temperature of water. The gas passes through the inner tube of a heat exchanger in the heat-recovery unit. The heat is then transferred to the cold water, which is flowing through the outer tube of the heat exchanger in the opposite direction. A tiny 30-watt circulating pump moves the cool water from the bottom of the water tank through the heat-recovery unit, and back into the top of the tank. By the time it returns to the tank, it is very hot. This process continues as long as the air conditioner is running.

Heat-recovery units are available from a number of manufacturers. The three companies listed below offer units that have been approved by independent testing laboratories for use in residential and commercial buildings. A letter to one or all of them will bring you complete information on their products:

Manufacturers of Heat-Recovery Units

Product	Manufacturer
Energy Conservation Unlimited, Inc. P.O. Box 585 Longwood, Fla. 32750	Energy Conservation Unit
G.S.T. Industries, Inc. 6605 Walton Way Tampa, Fla. 33610	Lectra Saver
Sun-Econ, Inc. Northway 10 Ushers Rd. Ballston Lake, N.Y. 12019	Econ-O-Mate

Windmills: An Overblown Prospect

Notwithstanding the considerable publicity given to windmills, they are not economical sources of electricity for private homes. The machinery is complex, the storage batteries are very expensive, and the amount of electricity you can get for the large investment is minuscule. They are romantic, inspiring, and good for pumping water, but at this point they are a poor substitute for the utility company.

SECTION 8

How to Buy Solar: Or Do You Need It?

Most people who had never heard of solar energy five years ago have now seen various diagrams of collectors, as well as pictures of happy solar customers, printed in their local newspapers; and they have read encouraging statements like this one: "Enough sun power falls on your house lot each week to heat the house for the entire winter." The only remaining questions are: (1) Is solar energy for you? And, if so, (2) When?

We will attempt to answer the "when" question first, because if the conclusion is "Later," then there is no point in thinking about it now.

There are differences of opinion, and you should know all sides of the debate. Since solar *water* heaters are slightly less complex, and more popular, than solar *space* heaters, we will consider the hot-water heaters first.

The Solar Water Heater

Reasons to Buy Now

(1) The quicker you install your solar water heater, the sooner you will benefit from reduced electric or fuel bills. Every extra year that you wait is another $200 to $300 spent on heating water—money you would have saved if you had already gone solar.

(2) Copper, the main ingredient in the good solar collectors, is not likely to get any cheaper. Increased

demand for solar products may have the same effect on the solar industry as it has had on the insulation industry: higher prices and reduced supplies. Labor and production costs may also increase. If these things happen, then you may pay less for a solar water heater now than you will pay later.

(3) Though solar water-heating systems have improved a bit in their overall efficiency, it may be a long time before any new technology takes a leap into better performance. The flat-plate collector used in Florida forty years ago is still the tried-and-true method for heating water with the sun. The flat-plate collectors on the market now are about as good as they are going to get.

(4) Oil companies have already invested in solar water-heater manufacturing operations, and utilities are getting into the act, too. The industry may be manipulated to serve the big interests. Sun-power equipment may some day be "rented" from utilities or oil companies, much as the phone company rents the use of its equipment. If this happens, the structure of the industry will change, and people's ability to buy their own solar heaters may be threatened.

(5) The income-tax breaks proposed by the federal government (see Section 11) for solar installations in private homes would be diminished every two years. That is, if you buy now, you could take more off your taxes than if you bought, say, in 1982. The tax bill was written this way to encourage an immediate and widespread conversion to solar energy.

Reasons to Wait

(1) Solar collectors are passing through a stage of extravagant infancy, similar to the early history of the television industry. They are being manufactured on a small scale, for a tiny, wealthy portion of the American market. When large-scale production begins, the prices may come down. An economic analysis done on solar-heating costs for the Energy Research and Development Administration assumed that current solar-heating costs were about $20 per square foot of

collector surface. The study also entertained the notion that those costs could be cut in half, to $10 per square foot, by 1980. Nobody can swear by such a reduction, but it is within the realm of possibility.

(2) State laws protecting "sun rights," and guarding the consumer against solar misrepresentation and fraud are just beginning to be enacted. As it stands now, there exists no nationwide protection for people who install solar collectors—only to have the sun cut off, unfortunately, by a neighboring construction project! Also, some fly-by-night companies have gotten into the solar business. The reliability of certain solar products has not yet been established. Methods of testing, evaluating, comparing, and certifying solar collectors are not yet fully developed. In a couple of years, the government will have put together a national-certification program, so people will know what companies to trust and what companies to ignore.

(3) Several states have passed laws offering tax incentives and property-tax exemptions on solar equipment. (See Section 11.) Additional state and local tax advantages may accrue to those who wait a couple of years before they go solar.

We could consider the pros and cons for a few more pages, but after weighing the arguments, our own conclusion is that there is no better time than the present. If you are convinced that a solar hot-water collector on your roof would be a good investment, and if you have already taken the low-cost steps to reduce your hot-water bill, as described in other sections of this book, then you have little reason not to go ahead. You get the extra federal-tax benefit by converting to solar now. Whatever additional incentives may be offered later will probably not equal the savings in electric costs you will achieve by going solar immediately. If the demand for solar hot-water heaters continues to increase, it is unlikely, in our opinion, that the price of them will go down. The technology for flat-plate collectors is far-enough advanced that you can already find a good one, provided you take certain precautions that we will describe later.

"Is It for Me?"

That leaves the "Is it for me?" question. In general, solar heating is for everybody. The copper collectors have proved their usefulness and durability in Florida, where they have been heating water for forty years or more. More than forty-five thousand solar units were being used in the city of Miami in the late 1940s, and in some sections of the city you can still find a solar collector on the roof of every house. A few years ago, the solar industry was limited to Florida and California, but now you can probably find solar heaters of some kind working in every area of the country. There are solar water heaters in Maine; and, if they work in Maine, they ought to work at all points south. The only area of dubious applicability is the Pacific Northwest, where cloudy days and low electric rates make the idea currently impractical.

In specific terms, "Is it for me?" can only be answered on a case-by-case basis. Just because your neighbor has a solar hot-water heater doesn't necessarily mean that the same thing will work for you. It depends upon whether your roof is strong enough to support the heavy collector, whether your house or apartment building has a southern exposure, and whether the sun has a clear path to your property. It depends, too, on whether local building codes and regulations allow solar heaters. These things you can discover by examining your home and by making a few phone calls.

The most important—and trickiest—calculation to make is the economic one. If you don't mind spending a little money to find out exactly how much you can stand to gain—or lose—from solar hot water, you can contact SOLCOST, a service of the Energy Research and Development Administration (ERDA). From SOLCOST you receive a questionnaire that you complete, describing your location, the design of your house, and your family's hot water needs, and send it to ERDA. They will send back to you a report on the size of solar collector you will need, the installed cost

How to Buy Solar: Or Do You Need It? 159

of the hot-water system, and a cost comparison with the system you now have. SOLCOST was provided free to people who applied for the ten thousand solar hot-water-heater grants available last year; but for everybody else, now, it costs $35. The address for SOLCOST is provided in Section 11. It's a terrific way to decide whether to buy a solar heater. But if you don't want to spend the money, you can proceed through the following calculations.

Electrically produced hot water accounts for about 20 percent of the average electric bill. A family can easily use 5,000 KWH a year to heat water—which represents $250 in electric costs now, at 5 cents per KWH, and up to $500 a year if rates double in a decade. Manufactured solar hot-water systems cost between $800 and $1,500, plus installation. On the face of it, it appears that solar hot water would pay back the investment within four to six years and provide free hot water from then on. The catch is that solar hot-water systems are not likely to provide 100 percent of your hot-water needs, because you would need too much collector surface and hot-water storage capacity to carry you over four or five cloudy days. A balance must be struck between the size and cost of the system and the percentage of hot-water requirements that it will cover.

In the southern and southwestern parts of the country, where solar hot-water heaters are a solid bet, the good ones can cut between 85 and 95 percent off a family's annual hot-water bill. In the Northeast, solar hot-water heaters were expected to provide 50 to 60 percent of the hot water, but a report by the A. D. Little Company suggests otherwise. Their interim report on a series of solar water-heating installations in New England says that results have not met expectations. It describes many problems in design, installation, and performance, and concludes that energy savings have been much lower than anticipated. The worst of the solar units have averaged less than a 5-percent savings; the best ones have cut only 37 percent off hot water bills.

So, how do you decide if your solar hot-water will

be a blue-chip stock or a losing poker hand? Various people have come up with complicated formulas, but we have two suggestions, depending on your financial situation: (1) If you can afford to buy the system outright, you will want to know whether the savings from solar use will exceed or fall short of the interest you could have earned by putting the money in a bank or in stocks or bonds; (2) if you buy the equipment on a time-payment basis, you will want to know whether the monthly reduction in electric bills covers the monthly payments you must make on the solar debt. These calculations will require a little wandering through the economic thicket, but they will help you make a good investment, or keep you away from a bad one.

The first thing to do is figure out how much you spend for heating water. In regions where solar heating works the best, water-heating expenses are the least. In places like Florida, the sun has already heated up the water in that natural solar collector, the ground. The electric or oil heater, then, has a lot less work to do in Florida than it does in Maine, where the water starts out much colder.

If you have a gas heater, you can figure how much gas you use for heating water by reading the meter or by looking at the utility bills for a period when you used the gas primarily for water heating. You won't get an *exact* amount this way, but make your best estimate and multiply it out in order to compute how much gas you need for a year's worth of hot water.

If you have electrically heated water but no electric meter, assume that hot water comprises 20 percent of your total electric bill. If you do have a meter and have been analyzing your daily KWH consumption, you may already have a very accurate notion of how much electricity you use for heating water. With whatever information is available to you, decide on an annual estimate, in kilowatts.

Of course, whatever amount of money you pay for heating water now, you will pay much more later. For the sake of the calculations, assume that electric or gas rates will double in fifteen years. Take the amount you pay now, divide by two, and add those two figures together. That total is your adjusted average cost for hot water over the next fifteen years. If you pay 5 cents a KWH now, the adjusted figure to use in your computations is 7 1/2 cents a KWH. Multiply the annual amount of gas or electricity you require for heating water by the adjusted fifteen-year price, to get an average cost for hot water over the next fifteen years. For example, if you decide that you will be paying an average of $300 a year for hot water over the next fifteen years, then your total water-heating costs will be $4,500. This $4,500, then, is the amount that the solar hot-water heater must save if it is to pay off in fifteen years.

What Does a Solar Water Heater Do?

The most important factor to consider, before buying, is what percentage of your hot water a given solar heating system claims to produce. You can't always believe the manufacturer. Never buy a solar water-heating unit without first visiting somebody in your area who has already installed the exact same system. Find out if it works and if the savings are as great as predicted. The investment is large enough to be worth your time comparing all the locally available systems.

There are a variety of designs on the market, but they all must do two things.

First, they must collect the sun's energy and convert it into heat. This is accomplished in the collector, the box-like structure that is mounted on the roof, or in some cases on the ground. The collector has a glass or plastic cover that lets in the sun's heat but doesn't let too much of it back out. Directly under the glass you find a metal plate, painted black, and covered with some sort of metal tubing. Water runs through the metal tube, and, as it travels across the collector, it reaches very high temperatures. How hot this water actually gets depends on what kind of collector you buy and the rate at which the water passes through the tube. Come collectors are much more efficient than others, because they produce more hot water in a given amount of collector area. Since the price of a collector is largely dependent on its total area, it is important to find one that puts out the most heat for the least money. The crucial calculations are cost per square foot, and BTUs per square foot. You want the most BTUs for the least cost.

Currently there is no good source for such information, since the federal testing program for solar collectors has not been completed. The Florida Solar Energy Center at Cape Canaveral does test the collectors of those manufacturers who send them in on a voluntary basis. The center has published a list of about thirty such collectors, including their characteristics and heat output. Most of these collectors are manufactured by Florida firms. If you are interested in finding the best ones tested, you can get a copy of the list by writing the center. Also, your state energy office probably has a directory of all the manufacturers and installers doing business in your state. You can check the efficiency claims made by the various companies and verify them with people who have used the systems. Dr. E. A. Farber, Director of the University of Florida solar lab, also suggests that you look for a collector with a certification from the American Society of Heating, Refrigeration and Electrical Engineers (ASHRAE). ASHRAE tries to test the durability of some collectors. Just because a collector does not have an ASHRAE certification doesn't mean it is defective, but you can be fairly certain that a collector that does have one will function as advertised, Dr. Farber says.

The second thing that a solar water-heating system must do is store all the heated water in some sort of insulated tank. If the storage tank can be mounted *higher* than the solar collector plates, the hot water will flow into it automatically. If the tank must be placed *lower* than the collectors, you will need to install a small circulating pump. A 30-watt pump will do the job, and it will cost about a penny a day to operate.

In the South, where water doesn't normally freeze at night, the water that comes from the solar collector can be piped directly into your home hot-water system and you can use your existing water heater for a backup. When the solar unit doesn't produce enough hot water, the electric element in your hot-water heater will kick on. In cold climates, the arrangement is a little more complicated. The water that runs through the collector must be protected by antifreeze, much like the water in a car radiator. Naturally, this water cannot be piped directly to your hot-water tank. The heat from the antifreeze solution must be transferred to your regular hot-water system by a device called a "heat exchanger." A heat exchanger isn't too big a deal, but it does add some additional cost and some extra potential problems to solar-heater installations in cold climates.

Comparing a Solar Water Heater's Likely Benefits with Its Cost

Once you have estimated your hot-water costs, and have decided *how much* solar collector you will require (a decision that local solar dealers can help you make), you can arrive at a fairly good economic judgment, the kind that SOLCOST would have provided you. You know how much your family will pay for heating water over fifteen years; in our example, it was $4,500. Let's say the solar company that you contact claims that their system will produce 70 percent of that hot water. A solar heater would then save you 70 percent of $4,500, or $3,150 over the fifteen years.

From this point, it is simple to compare the *benefits* with the *cost*. The cost will depend on what you actually buy, but, for our theoretical case, let's say that the solar water heater, installed, costs $1,600. And don't forget to figure maintenance in your balance sheet; maintenance for a well-constructed system should not be too great, but, for the example, we'll assume that you pay $200 over fifteen years for maintenance. The total cost of the solar heater, then, would be $1,600, plus $200; or $1,800.

So far, we have a savings of $3,150 measured against an investment of $1,800. But there are two additional factors that must be considered: The first is tax incentives. The federal government now offers tax breaks to homeowners who install solar hot-water or home-heating apparatuses. Many states, too, have their own tax incentives. You must look into this for yourself, but let's assume, for this example, that you can cut $150 a year off taxes, for a five-year period. This makes solar a little more attractive—but you must weigh the benefit against whatever tax disincentives exist in your state. Be sure to find out if you are liable for additional property taxes because of your solar collector. Many of the early solar converts ran into high property taxes that made their efforts much less rewarding than they had earlier assumed. But since many states now exempt solar installations from property tax, we will assume that none will be levied on our theoretical solar heater. The solar balance sheet now should look like this:

Benefits	Costs
$3,150 savings in water heating costs (provides 70% of needs)	$1,600 for solar heater
$ 750 state and federal tax break ($150× 5 years)	$ 200 for maintenance
$3,900	$1,800

From this benefit/cost chart, we can easily find the break-even point. We are saving $3,900 over fifteen years—an average of $260 each year. It will take between six and seven years to make back the $1,800 investment, and after that we will continue to benefit at the rate of at least $260 a year over the life of the system. If the solar heater lasts fifteen years—and good ones should work for much longer—then our net gain after we reach the break-even point would be a minimum of $2,080.

Now, the way to make your *final* decision—if you buy the system outright—is to calculate whether that gain would be greater or less than the interest you would earn if you just put the $1,800 in the bank, or invested it otherwise.

Six-percent interest would be a conservative return on an investment, but we will use that figure. At 6 percent, the original $1,800 would earn $2,512 over the fifteen-year span, whereas the gain from the solar investment would be $2,080. This makes it sound more attractive to put the money in the bank—but only before we consider the income taxes on interest earned! If you are in a high tax bracket, the government would take enough of the interest accrued on the $1,800 investment so that you would be better off to buy the solar heater. If you are in a low tax bracket, the decision, in our theoretical example, would be an economic toss-up. But remember that solar heaters will probably last longer than fifteen years, and will give you a hedge against a precipitous rise in electric costs, against a power shortage, and against mandatory power rationing. Furthermore, the solar installation, for all we know, may last forty years, and may be a big selling point if you decide to put your house on the market. These are factors that only you can evaluate. The important thing is not to overlook any important elements in your decision: the cost of heating water; the percentage of your hot water the solar system will provide; the possible additional property tax; the cost of the solar heater and its maintenance; and federal or state tax breaks.

Borrowing for Solar

You may not be able to, or wish to, pay for the solar water heater out-of-pocket—if you decide to buy one. If you borrow the money, you will have to include the costs of a loan in your calculation. In this case, we think it is more useful to base your decision on a short-term analysis. Figure your water-heating costs on a monthly basis, and at the same adjusted electric rate we described earlier. The rate to use for this calculation is halfway between what you pay now and what you will be paying in ten to fifteen years.

Assume that rates will double; so if you pay 5 cents per KWH now, use 7.5 cents for your computations.

Talk to the manufacturer's representative and to people who have installed the system, in order to decide what percentage of your hot-water needs the system will produce. You can then decide how much money the solar unit will save you each month. If the payback period for the loan is equal to the assumed life of the solar heater (we are assuming only fifteen years), then the decision is easy. If the monthly loan costs, with interest, are less than the monthly savings on heating water, then the solar heater is a good deal. If the monthly loan costs exceed the savings on heating water, then solar is not a good investment. (For loans with shorter payback periods, the computations are more complex, and you will have to make your own analysis.)

Final Advice

Whatever you do, it is important that you make careful financial calculations. The solar-heater people or the banks will probably help you. But even if the bottom line favors the solar heater, here are the steps you should take to protect yourself:

(1) Don't buy a solar hot-water heater without first checking with somebody else who has purchased the same kind of unit.

(2) Make sure that the installation is guaranteed, and that any mistakes in setting up your system will be paid for by the solar company in question.

(3) See if your state has a certification program, and try to buy a unit that has been certified.

(4) Make certain the company sends out a qualified installer. Solar installation is a tricky matter, involving plumbing, electricity, carpentry, roof repair, and solar physics. If the collectors are installed at the wrong angle, or pointed in the wrong direction, the value of the solar system will be severely impaired.

(5) Make sure nobody is planning to put a house or a tree in front of the collectors that you are putting on your roof.

(6) When in doubt about the merits of one collector over another, choose copper. Copper solar collector plates and tubing have lasted for more than forty years in the salty, humid Miami air. Plastics are less satisfactory because they are partial insulators, and because they tend to deteriorate when exposed to the sun for long periods of time. Aluminum collectors are sometimes less expensive to buy, but their longevity has yet to be fully established. A close runner-up to copper is the bonded, expanded-metal collector plate. (In this process, two sheets of metal have been fused together, then expanded to provide an internal passage for the liquid to be heated. A major advantage to this system is that it lends itself to mass production. The disadvantage is that, because of possible corrosion in the metal, special fluids must be used in the collectors, making the use of the heat exchanger mandatory.)

A few of the highly efficient collectors on the market are part of the "second generation"; they have retained the basic elements of the flat-plate collector, but have been modified to improve their performance. One such is produced by Daystar, in Burlington, Massachusetts, and uses a patented heat trap that creates much higher temperatures around the inside of the collector box. Another is a vacuum-tube solar collector produced by General Electric Company. It can trap up to twice the amount of energy collected in a standard model.

Very high temperatures are not necessary for simple solar water heaters, but they are indispensable in projects that combine both solar heating and solar air conditioning. You can keep up with such innovations through your state energy office, or through the National Solar Heating and Cooling Information Center, P.O. Box 1607, Rockville, Maryland 20850.

Doing It Yourself

One way to substantially cut the costs of a solar hot-water heater—if you decide it *is* for you—is to build it yourself. *Popular Science* magazine often publishes articles about successful collectors that were designed and built by backyard amateurs. Many of these designs work well.

Building a solar heater from scratch can be an exciting and rewarding part-time project, but you must possess a tinkerer's temperament and some basic handyman skills. You have to be able to solder tubing together, build wooden collector boxes, install the unit on the roof without damaging the roof or creating leaks, and connect the whole thing to your plumbing system without giving the toilet a hot-water flush. Many of the plans published in magazines are not detailed enough to provide answers to all the questions that come up along the way. Some of the advertised "how-to" books really don't tell enough. One of the authors of this book tried to adapt a solar design from a magazine—and was left with a lot of cut-up pipes and a deficit of $200!

On the other hand, some step-by-step solar-heater plans work well, and don't assume prior knowledge. You of course need to know how to build wooden boxes and how to solder tubing, but from then on the process is easy. One of the best plans (see Plate 11) is offered by the Florida Conservation Foundation, Inc., 935 Orange Ave., Suite E, Winter Park, Florida 32789. The twenty-four-page booklet, called "How to Build a Solar Water Heater" really tells you all you need to know, and costs $2.00, plus 50 cents postage. The collector you build is the copper type that has been popular in Florida since the 1940s. The costs of construction and installation run between $800 and $1,000. The other author of this book built one of these solar heaters two and a half years ago, and it has worked flawlessly ever since. The system supplies him with 99 percent of his domestic hot water during the summer and about 95 percent during the winter. (If you live in a colder climate, you will have to adapt a heat exchanger to the solar unit, but heat exchangers are not difficult to install.)

The Solar Space Heater

The solar *water* heater is the more widely used solar device, but plenty of people are now also heating their *homes* with the sun. Solar space heaters, as they are called, work on the same principle as the water heaters. In fact, it often makes sense to combine the two in a single system.

The many varieties of solar space heaters can be divided into two categories.

First are the liquid systems. These employ flat-plate collectors that are identical to those used for solar water heating, except that they must have a much larger area. The liquid that flows through the collector, and is heated by the sun, is piped into large, insulated storage tanks, which can store the heat for days. That heated liquid is then either run directly through radiators, or is sent into a heat exchanger, which removes the heat from the liquid and blows it through the ducts of a forced-air furnace.

The second category of solar space heaters uses air instead of a liquid. Air is forced through the boxes on the roof, and the air itself is heated by the sun and then blown through heating ducts. The hot air manufactured during the day is collected in a large rock pit, where much of the heat can be retained and used during the night. Air collectors are a little less complex than liquid collectors, since no antifreeze solutions or heat exchangers are necessary. But air collectors do require a large pit area of several tons of rock for heat storage. Solar air systems also use massive air ducts, and sometimes demand an elaborate network of flues, dampers, and blowers.

"Is It for Me?"

The economic factors that you must consider in deciding on a solar space heater are the same as the ones used in deciding on the water heater. You must know how much you already pay for fuel or electricity for heating, and you must make an educated guess as to how much these prices will go up in the future. You can then estimate the percentage of your heating load that will be absorbed, or provided, by the solar unit, in order to decide if the savings will exceed the costs of the solar product. Solar-space-heater installations are considerably more expensive than solar water heaters, and the difficulties in adding them to an existing house can be numerous. Solar space heating is a modification that will apply to only a small percentage of homes that already exist. Depending on the type of solar heater that you choose, solar space heating can cost anywhere from $5,000 to $10,000. You will have to check with local sources to decide if solar space heating would benefit you, but ERDA has published a booklet that will help give you a general idea. The booklet divides the country into two climate zones, as shown in Plate 12.

ERDA then estimates the size of the solar collector and storage tank required to heat a 1,500-square-foot home in these various zones (Table 10).

And finally, the booklet estimates the annual dollar savings from the installations (which are assumed to provide hot water as well as home heating) (Table 11).

If you want a copy of the booklet, called "Solar Energy for Space Heating and Hot Water," it's available from the Government Printing Office, Washington, D.C. 20402, for 35 cents. It is a useful summary of the types of solar heaters available, and of the general economics of solar heating.

ERDA did a further analysis of the prospects for solar heating in various parts of the country ("An Economic Analysis of Solar Water and Space Heating," November 1976). They considered two ways to measure the cost effectiveness of solar installations: positive savings, or the year in which the savings on fuel money exceeds the annual costs of the solar unit; and payback, or the year in which the net savings off the fuel bill equals the remaining principal on the mortgage

Table 10: Approximate Collector and Storage-Tank Sizes Required to Provide the Heating and Hot Water Needs of a 1,500-Square-Foot Home

Climate Zone	Percent of energy supplied by solar	Collector Area (sq. ft.)	Representative Collector Dimensions		Storage Tank Capacity (gal.)	Representative Cylindrical Storage Tank Dimensions	
			No. of 8' high row	Length of each row (ft.)		diameter (in.)	Length (in.)
1	71	800	3	33	1,500	48	200
2	72	500	2	31	750	42	138
3	66	800	3	33	1,500	48	200
4	73	300	1	37.5	500	48	78
5	75	200	1	25	280	42	60
6	70	750	3	31	1,500	48	200
7	70	500	2	31	750	42	138
8	71	200	1	25	280	42	60
9	72	600	2	37.5	1,000	48	132
10	58	500	2	31	750	42	138
11	85	200	1	25	280	42	60
12*	85	45	1	5.5	80	20	63

Table 11: Energy Supplied and Annual Dollar Savings (1,500-Square-Foot Buildings)

Climatic zones	Energy savings, millions of BTUs	Comparison with oil		Comparison with electricity		
		Equivalent gallons	Dollar savings *	Equivalent kilowatt hours	Dollar savings at indicated cost	Cost per kWh
1	67.9	757	303	19,900	995	5¢
2	54.9	612	245	16,000	720	4.5¢
3	82.0	914	366	24,000	960	4.5¢
4	41.8	466	186	12,200	488	4¢
5	33.8	377	151	9,900	347	3.5¢
6	98.9	1,103	441	29,000	1,015	3.5¢
7	50.6	564	226	14,800	518	3.5¢
8	39.0	435	174	11,400	399	3.5¢
9	74.6	832	333	21,900	767	3.5¢
10	46.5	518	207	13,600	272	2¢
11	43.7	487	195	12,800	512	4¢
12	16.7	186	74	4,900	196	4¢

* 65% furnace efficiency at 40¢/gallon.

Economics of Solar Hot Water and Heat [1]

	Years to positive savings				Years to payback			
	Elect.	H.P.	Oil	Gas	Elect.	H.P.	Oil	Gas
ATLANTA	3	*	*	*	14	*	*	*
BISMARCK	4	7	7	*	14	19	18	*
BOSTON	3	*	*	*	14	*	*	*
CHARLESTON	1	*	7	*	11	*	19	*
COLUMBIA	3	*	*	*	14	*	*	*
DALLAS/FT. WORTH	3	*	*	*	13	*	*	*
GRAND JUNCTION	1	6	6	*	12	17	17	*
LOS ANGELES	1	7	7	*	10	19	19	*
MADISON	3	*	*	*	14	*	*	*
MIAMI	1	8	7	*	9	20	19	*
NEW YORK CITY	1	*	*	*	12	*	*	*
SEATTLE	*	*	*	*	*	*	*	*
WASHINGTON, D.C.	4	*	*	*	14	*	*	*

[1] The figures in this chart are for a solar unit providing 60 percent of home heating load, where the cost of solar installation does not exceed $20 per square of collector. It also assumes that the cost of fuel will rise 10 per cent each year.

for the solar system. "Positive savings," ERDA says, "can be viewed as similar to 'dividends' paid on investment."

In ERDA's calculations, a solar heating system is economical if positive savings occur in five years or less, or payback occurs in fifteen years or less. In a thirteen-city analysis, ERDA figured the solar savings in comparison to four conventional methods of heating: electricity, heat pumps, oil, and gas. Their conclusions are contained in the following chart. (An asterisk means that the solar-heating load is insufficient to justify the expense of a solar installation at this time.)

You can use these numbers to help you decide whether solar space heating is worth investigation, but you will have to do your own figuring before you can make an intelligent evaluation. (For more information on solar loans, see Section 11.) The best way to find out about the various companies that make or install solar space heaters in your area is to write the state energy office.

Do-It-Yourself Solar Space Heating

It is more difficult to build and install a solar space heater, or a space heater and water heater combination, than it is to build a solar water heater by itself. Nevertheless, the National Aeronautics and Space Administration (NASA) puts out a do-it-yourself plan that is very easy to understand and has tested out well. You can get a copy for $4.25 by writing the National Technical Information Service, U.S. Department of Commerce, Springfield, Virginia 22161. The plan includes collectors, a heat exchanger, water pump, storage tank, piping, and controls to make the system completely automatic; and the whole thing can be put together for $2,000, or one-third the cost of some of the manufactured systems. Tests indicate that a homeowner can reduce his heat bill by approximately 40 percent for a 1,500-square-foot home, insulated to 1974 FHA minimum standards. Payback is estimated to occur in ten years.

Swimming-Pool Solar Heaters

Heating a swimming pool with oil or electricity is the kind of extravagant practice that may someday be outlawed. When cities in California had to make quick energy cutbacks, they immediately moved to turn off their swimming-pool heaters. So, even if you can afford one, you may not always be able to use it. Solar heating for swimming pools is easy to install. The water only has to be raised a few degrees, instead of the tens of degrees necessary for producing domestic hot water. The collectors and connecting pipes don't have to be insulated, and they don't have to withstand high water pressure. For that reason, swimming-pool collectors are relatively inexpensive. In addition, no holding tank is required; the water just keeps flowing through the collector and back into the pool.

You can make a cheap solar pool heater with a large piece of black plastic (or other products, described in Section 4), but if you want to go the full route, solar collectors are available. A number of companies that manufacture collectors suitable for use as swimming-pool heaters are listed at the end of this section.

Solar Air Conditioners

It will soon be practical to cool houses with solar energy. The paradox is that to produce sufficient cooling power, you have to generate very hot liquids in the solar collectors. The earlier collectors could not reach the high temperatures required, but the newer ones, such as the Daystar Collector, can do it; once temperatures in excess of 180° are reached, an absorption-type cooling system can be used. It works the same way as the gas refrigerators used in modern recreational vehicles. A number of experimental solar cooling systems are now in operation around the country.

A picture of the first building in New England to have solar heating *and* cooling is shown in Plate 13.

The building is the Highland Avenue Branch of the Concord, New Hampshire, National Bank. We visited the building, and the solar unit seems to be working well. It provides between 50 and 60 percent of the heating and cooling needs for the bank—a remarkable achievement, since 50 percent of the wall surface of the building is glass. The system uses twenty-six Daystar solar collectors and an Arkla absorption-cycle air-conditioning unit. Such installations are not yet economically attractive for homeowners. However, William G. Hillner, Commercial Sales Representative for Daystar Corporation, predicts: "It will become very practical when the cost of oil and gas increases in price from the present levels."

A Solar Catalogue and Suggestions for Further Reading

Until recently, people who built their own solar water heaters had a hard time finding a centralized source for all the pumps, controls, tubing, and storage tanks that such installations require. Nor did people who just wanted to buy an already-manufactured solar product know where to look. But at least one solar clearinghouse is now in operation.

Solar Usage Now publishes a 127-page catalogue that lists nearly every conceivable solar-related product available. What's more, they provide a mail-order service and will ship any item they describe in the catalogue. The catalogue includes thirty pages on different solar collectors and systems, not to mention separate listings for liquid storage tanks, special pipe fittings, heat exchangers, precision thermometers, pipe insulation, solar controls, etc. You can get the catalogue by sending $2 to Solar Usage Now, Box 306, Bascom, Ohio 44809.

Popular Science has published more material on solar innovations than any other popular magazine. They have combined a lot of this material into a "Solar Energy Handbook," a compendium of laws and government incentives, economic calculations, technical descriptions, and plans for how to install various

How to Buy Solar: Or Do You Need It?

types of solar heaters. The handbook is available from the magazine, Plans Dept. SE, 380 Madison Ave., New York, N.Y. 10017, for $1.95.

Solar-Product Manufacturers

Many long lists of solar manufacturers are printed in government publications and magazines. We picked up one such list of two hundred manufacturers, and wrote each of them a letter. Enough of the letters were returned with "Moved, Left No Address" stamped on them that we knew the solar-heating business includes more than a few flaky enterprises. That's one of the reasons it is important to investigate before buying: if the company you deal with goes out of business, they won't be around when and if problems arise with your system.

Many companies did not answer our letters, but the ones that did are listed below. We have also included a short description of the products they make.

Manufacturers of Solar-Related Products

Manufacturer	Solar-related products
Aluminum Company of America ALCOA Building Pittsburgh, Pa. 15219	Solar pool heating system; solar heating system controls
AMETEK Power Systems Group 1 Spring Ave. Hatfield, Pa. 19440	High-performance solar collectors
Chamberlain Manufacturing Corp. 845 Larch Ave. Elmhurst, Ill. 60126	Chamberlain flat-plate collectors for domestic hot water and space heating
CSI Solar Systems Division 12400 49th St. S. Clearwater, Florida 33520	Sol-Heet collectors for domestic hot water, space heating and swimming pools
Daystar Corporation 90 Cambridge St. Burlington, Md. 01903	High-efficiency solar collectors for heating and cooling

Manufacturer	Solar-related products
E & K Service Co. 16824 74th N.E. Bothell, Wash. 98011	SOL-R-Panel for domestic hot-water heaters, space heating, and swimming-pool heating
Energy Systems, Inc. 4570 Alvarado Cyn Rd. San Diego, Cal. 92120	Solar collectors for domestic hot water, space heating, and swimming pools
FAFCO, Inc. 235 Constitution Dr. Menlo Park, Cal. 94025	Pool-heating sysetms, automatic-control sysetms, and heat exchangers
General Electric P.O. Box 13601 Philadelphia, Pa. 19101	High-efficiency space-heating and cooling systems, using vacuum process
Grumman Energy Systems 4175 Veterans Memorial Highway Ronkonkoma, N.Y. 11779	Solar collectors for domestic water systems, space heating, swimming pools, and space heating (air distribution)
Inter Technology Solar Corp. 100 Main St. Warrenton, Va. 22186	Solar collectors; domestic hot water kits
Ilse Engineering, Inc. 7177 Arrowhead Rd. Duluth, Minn. 55811	Sandwich solar absorber panels for domestic water heating and space heating
Kalwall Corporation P.O. Box 237 Manchester, N.H. 03105	Solar air heater (space heating); movable insulation; reflective curtain; Sun-Lite glass fiber for solar-collector glazing
KEN-TEC Corporation 5309 Shirley St. Naples, Fla. 33941	Solar collectors for domestic water systems, space heating, swimming pools
L O F Solar Energy Systems 1701 East Broadway Toledo, O. 43605	Sun Panel solar collectors for hot-water and space-heating and cooling systems
Olin Brass East Alton, Ill. 62024	Component supplier to solar industry; Roll-Bond solar products; copper or aluminum panels, bonded and expanded
PSG Industries, Inc. 1225 Tunnel Rd. Perkasie, Pa. 18944	Accustat solid-state wall thermostat, with preset temperature sensor
Raypak 31111 Agoura Rd. Westlake Village, Cal. 91359	Solar-Pak collectors for water-heating systems and swimming-pool heaters

How to Buy Solar: Or Do You Need It?

Manufacturer	Solar-related products
Reynolds Metals Company Richmond, Va. 23261	Solar flat-plate collectors, using integrally finned extruded-aluminum tube
Rho Sigma, Inc. 11922 Valerio St. North Hollywood, Cal. 91605	Solar controls and instrumentation
Solar Energy Research Corp. 701 B South Main St. Longmont, Colo. 80501	Kits for domestic hot-water systems, solar-space-heating and -cooling systems, solar pool heaters, storage units, and pumps
Solaron Corporation 300 Galleria Tower 720 S. Colorado Blvd. Denver, Colo. 80222	Solar air-heating systems; solar domestic hot-water units
Solar Systems, Inc. 507 West Elm St. Tyler, Tex. 75702	Solarvak Flat Plate Solar Collector
Sundu Company 3319 Keys Lane Anaheim, Cal. 92804	Solar swimming-pool heaters
Sunwater Energy Products 1488 Pioneer Way El Cajun, Cal. 92020	Solar collector panels for domestic hot water, heating, and cooling; also solar water-purification systems
Tranter, Inc. 735 East Hazel St. Lansing, Mich. 48909	Solar collector absorber plates; storage-tank heat-transfer coils; solar heat exchangers
Vapor Megatherm 803 Taunton Ave. East Providence, R.I. 02914	Thermal storage systems (used as supplemental system on solar applications)
Walker & Mart Laboratory, Inc. 633 9th St. N. Naples, Fla. 33940	High-efficiency energy-collecting and -storage unit, using liquid or salt as storage vehicle
Zomeworks Corporation P.O. Box 712 Albuquerque, N.M. 87103	Solar water heater; Skylid, Beadwall, and Nightwall for home heating; freon gravity drives and sun calculators

SECTION 9

Energy Survival: What to Do Before the Power Fails

New York City recently went through a one-day blackout, and some people predict that the fuses will blow in other parts of the country as power demand increases and supply diminishes. Whether or not you live in a blackout-prone area depends on the quality of your local utility and the reliability of its power sources. Good information on the probability of blackouts is hard to find. Consolidated Edison, the New York electric utility company, had just finished reassuring its customers that another blackout—like the one in 1965—couldn't happen, when it did.

One thing that struck people in New York was how dependent they were on the electrical "umbilical cord." You can, of course, simulate their experience by pulling the main switch on your house and stumbling around for the next twenty-four hours in an energy drill. But even that simulation won't reflect what New York went through. Nobody knew when the power would come back. Many people were amused to discover how much they relied on a plethora of electric gadgets; they had misplaced the manual can openers years ago, and they couldn't even write letters of complaint on their electric typewriters!

The thing that was *not* amusing was the shortage of water. Apartment dwellers rely on electric pumps to bring water up to their kitchens and bathrooms. The pumps stopped working, and they had no way to flush toilets, take showers, or wash dishes. These drought-like conditions lasted only a few uncomfortable hours, but they point to the wisdom of preparing for blackouts. A prolonged power failure could not only be inconvenient, but also unhealthy.

The energy-saving measures described throughout this book would take on added importance in a blackout situation. Some of the things you have done to conserve energy should also help you get through periods when there is none. If you have bought water-saving toilets or showerheads, you will be better able to cope with water shortages. If you have installed a solar space heater or a wood-burning stove, you will be prepared for winter fuel shortages. If you have devised an alternate cooking source, you can at least prepare a meal when the lights go out. But, in addition to what you may have done already, here are some specific suggestions to make a blackout—should one occur—less troublesome:

(1) *Discover your water.* At the first news of a blackout, people are told to fill their bathtubs and other available containers with water. Often, in large buildings, the rooftop tank will hold water pressure for a couple of hours, but it is best to stock up while you can. Such last-minute measures are usually sufficient to get people through a short power failure, but they won't work for an extended blackout. You may be able to avoid a real water crisis by discovering where your home or building actually gets its water, how much is available if the power goes out, and whether your water system depends on electricity for distribution. Furthermore, find out if there is another way for you to obtain water. Are there alternative sources, like the fire hydrant down the street, or the hot-water tank in the basement? (Homeowners sometimes forget that they have their own 40- to 80-gallon reserve, which can be obtained by opening the tap at the bottom of the hot-water tank.)

Citizens who live in large apartment buildings might benefit from a generator large enough to run the electric water pump, if they haven't one already. In New York, there was still sufficient pressure for water to flow through the city mains, but not enough to push the water to the upper floors of buildings. Buildings that have small generators could continue to provide water, thus eliminating the major inconvenience of a blackout. A 6.5-kilowatt generator for that purpose would cost about $2,000. If you live in an apartment building, you might want to work with your tenants' association on a generator project. Find out if your city could still deliver water to the building during a power failure; if it could, see if you can induce the owner(s) to invest in a generator.

And there are ways to get by with *less* water. Assuming you can do without the daily shower, the critical need for water is in washing dishes and flushing toilets. The dishwashing problem is even largely eliminated if you have bought an emergency set of plastic forks and spoons, plastic glasses, and paper plates. The toilet problem is more difficult, and if you can't find a reliable source of emergency water in your area, you might consider buying a marine toilet. Perhaps two or three families could buy one together. A marine toilet uses a minuscule amount of water for each flush. It can be useful not only during power failures, but also during droughts—when water conservation is mandated by law. Some people even believe that marine-type toilets should be used in normal non-emergency circumstances, since conventional toilets require 5 gallons of precious fresh water per flush. Several varieties of marine toilets are available through stores that carry boating or camping equipment; they can cost anywhere from $75 to $400. If you use a marine toilet in an apartment, you will probably also have to obtain some kind of holding tank. The solid waste is liquefied in the holding tank, so that it can be flushed out through the apartment sewage system later. *New York* magazine, in a recent article on how to survive a blackout, emphasized the importance of the marine toilet.

(2) *Get a Styrofoam cooler*. The second major problem in a blackout is refrigeration. Most food spoils rapidly, and frozen foods thaw quickly, especially in the summer months, when blackouts tend to occur. If you have a freezer, or even if you only have a freezer compartment in your refrigerator, try to open

it as little as possible after the power goes off. In fact, you can keep it securely closed by wrapping duct tape around the edge of the door. If you do have to open your refrigerator or freezer, remove everything that you need at one time; eat or drink the perishable items first.

The blackout offered compelling proof to most New Yorkers that their refrigerators and freezers are not well-insulated. Ice cream began to melt after a few hours. Frozen foods were spoiled in less than twenty-four hours. It may be difficult to believe, but some of the Styrofoam boat coolers sold at marine stores are better insulated than the expensive refrigerator in your apartment or house. These coolers, when packed with some ice, or even when stuffed with newspaper, will maintain frozen foods for as much as three days. Having such a cooler in your storeroom is a good blackout protection. You can fill it up with all your frozen food, wrap the outside edges with tape, and leave the box in the coolest spot in your home. The food will not only be preserved, but you can then open the regular freezer from time to time without fear of spoiling *all* of your groceries.

(3) *Have candles and flashlights handy*. The thing people talk about most during blackouts is how the lights went out, but lights were the least of New York's worries. Candles appeared all over town. As long as you remember to keep a supply of candles, or a couple of hurricane lamps, or a flashlight or two, then you will be in good shape.

With the marine toilet, the Styrofoam boat cooler, the candles, and a battery-powered radio or TV, living through a blackout becomes the residential equivalent of taking to the woods. If you are not prepared, a blackout can be unpleasant. If you are prepared, it can be a camping trip organized by your local utility company.

A few families have insured their partial self-reliance by the purchase of a small gasoline generator for their homes. The problem with generators is that they are expensive relative to the amount of power they actually produce. Some generators on the market cost only a few hundred dollars, but they supply merely enough electricity to burn a couple of lights or to run one large appliance. What's more, if you own a generator, it is necessary to run it for fifteen or thirty minutes every few weeks in order to keep it in good working order. Failure to do this will insure that the machine is inoperative when you need it.

The Survival Room

Most blackouts have occurred in the summer, which is a lucky thing for the people who have experienced them. But what would happen in a winter blackout? Or—even if there is no blackout—what if fuel oil or natural gas became scarce or nonexistent in some parts of the country—maybe yours?

One alternative to freezing in such situations is what Evan Powell, a writer for *Popular Science* magazine, calls the "survival room." He believes that people should be as energy-independent as possible, so that they can live through an energy crisis with a minimum of difficulty. A survival room, Powell says, is simply a well-insulated section of the house, where the entire family may sleep or coexist for a few days if power goes off or fuel runs out. Such a room might contain, first, a fireplace, or a small heater run on something other than the main fuel for the house, or one of the wood-burning stoves described in Section 5. It might also contain a campstove kitchen, some kerosene lanterns or candles, and, if the water supply is far away, a marine toilet and a small camper sink with its own water supply.

A survival room can be part of an existing house, possibly the family room. The important thing is that it must be designed so that it can be cut off from the larger living area—in order that it can be kept warm with a minimum of fuel. (The little heat escaping from this room would help, of course, to maintain a high enough temperature in the rest of the house to prevent pipes from freezing.) The room should contain a minimum of windows or outside openings, and it should be heavily insulated.

You don't need to wait for a fuel shortage or a power failure to use your survival room, although it would be especially helpful in those situations. During last winter, newspapers were filled with stories of people who couldn't keep up with high electric and fuel bills. They were forced to reduce thermostats to shivering levels, or in some cases to abandon their homes. If they had a survival room, on the other hand, they could stay at home, keep warm, and not go into debt. They could retreat to the room during the coldest periods of the winter, leaving the rest of the house at a minimum temperature to protect the structure from frost damage. The survival room would be used as kitchen, living room, and bedroom on those occasions, offering cheap heat and close companionship. Historically, there were times when families would routinely hole up in smaller spaces during severe winters. The survival room is a modern adaptation of that technique.

Some of the energy-saving products we have suggested in other sections of this book could be used in the survival room. The parlor stove, for instance, works as fireplace, as high-efficiency wood-burning heater, and as cooking surface.

SECTION 10

The Apartment Connection: Ways to Save Energy in Big Buildings

A lot of people don't have meters to read or houses to insulate, but they still pay big electric and gas and, sometimes, fuel-oil bills. They are the residents of apartments in large buildings. The plight of the apartment dweller goes unnoticed in most of the material we've presented on how to save energy. It is assumed that renters won't be around long enough to bother with faucet aerators and hot-water-tank insulation jackets, and since renters don't own apartments, they certainly won't want to pay for insulation or solar heating. Similarly, many people *own* apartments in large buildings. In their cases, the most notorious energy wasters—old, oversized water heaters, outdated, inefficient furnaces, etc.—belong to everybody in the building, so nothing can be improved without going through a cumbersome process of getting everybody to agree. And, most important, their utility costs are often prorated and spread out among all the tenants in the building, so that most of them don't have the foggiest idea of how much power they consume individually.

Even if we admit to apartment dwellers' and owners' peculiar situation, it may be even easier for tenants and owners of apartments to cut their utility bills than it is for owners of private homes. For one thing, there are a great many people in a big building. Large energy-conservation projects can sometimes sustain themselves better there than in a one-family house. Tenants' associations or co-op groups can hold energy huddles to report on progress, cheer up the doubters, and inspire the laggers. For another thing, gestures such as reducing the wattage of light bulbs—perhaps insignificant in a small home—can save hundreds of dollars a year in a large building. Those first and easiest steps in the energy plan, presented in our first chapters, can return inspirational dividends. And, if apartment owners or tenants do decide to do more, they can finance major improvements with less difficulty, and less per-person expenditure, than individual homeowners. They can, for example, invest in insulation at a lower price—because they are buying in quantity. A few energy-saving devices, such as the heat-recovery unit that produces free hot water from the central air conditioner, may be uneconomical for many single homes but highly profitable for large buildings.

If, then, you rent or own an apartment, you may want to consider the following energy-saving programs strategies:

Investigating Your Energy Rights as a Renter

Renters pay a heavy penalty for structural deficiencies in the buildings they occupy—especially in old, drafty apartments with inadequate furnace-heat delivery systems—for landlords merely tack the resulting high utility costs on to the rent without itemizing those costs. The landlord can pass the burden along; he, himself, has no incentive to conserve. Many landlords feel it would not be worth their effort to insulate or to refurbish a furnace in an old building that will be torn down in a few years, anyway.

Among all the questions that careful apartment shoppers ask about safety, cleanliness, the rent deposit, and whether cats are allowed, the apartment building's own energy bill is rarely mentioned. Yet that bill can be the primary reason why rent increases are demanded by the building owner, at least in cases where the owner pays for the utilities—heat, hot water, and sometimes gas for the kitchen stove and oven —out of his own pocket. If the owner doesn't pay the utilities, the renter can get into a deeper bind. His budgetary calculations can be thrown completely awry when he realizes that to survive in his uninsulated apartment, he must pay out hundreds of dollars for gas, or oil, or electric heat.

The legal aspects of renters' rights in buildings that waste energy have not been fully explored, but you can at least do some investigating on your own. Before the rental contract is signed, find out if the landlord has undertaken any of the energy-saving steps described in this book. Ask if the water tank is insulated, or even if flow restrictors are installed on the showerheads, and so on. Find out, if you can, what percentage of the rent you pay him goes toward utility bills. Perhaps it is possible that if you and the other tenants are willing to undertake a conservation program, the owner will reduce the rent by a percentage of the amount you can save him on utilities. Many people get rent rebates through an apartment-painting arrangement; energy conservation can perhaps provide a similar opportunity. In fact, energy conservation is one of the few tenant–owner projects that both sides could view positively.

(If you have not already chosen the apartment you want to live in, the unit you take in the building could have personal economic significance to you. A study done in Davis, California, found that some apartments could be much hotter, or cooler, than neighboring apartments in the same building, depending on which way they faced. The orientation of the apartment you choose may help determine the size of your air-conditioning or heating bill. The information in Section 13, which discusses how to design energy-saving houses, should also help you have an energy-saving apartment.)

Banding Together for Fun and Profit—in a Master-Metered Building or Co-op

Your individual efforts will take you only so far. Collective conservation programs in apartment buildings offer even greater benefits. In fact, the success of a few such programs contradicts the pessimism with which the U.S. government views energy-saving prospects in master-metered buildings. ("Master-metered" means that the entire electric service for a building is computed on one single meter. It is estimated that one-third of all apartment buildings in the country are master-metered for electricity, and 85 percent of apartment buildings that use natural gas operate on a master meter.)

The government has tagged master meters as an energy menace. Since people are not billed directly for the power they use, the argument goes, then they have no incentive to conserve. Why should one person turn off an air conditioner and sweat stoically away, while the neighbors are re-creating Antarctica in their living room? This kind of comparison occurs to everybody, and so the four million people who live on master meters are said to use 20 to 30 percent more electricity per capita than people who reckon with their individual meters. Because of the master meter, tenants may pay more than $200 a year in unnecessary electric bills.

The Carter Administration would like to do away with this gluttony by prohibiting master meters in most new buildings. The problem is what to do in the old buildings that already have them. Utility companies talk about converting the buildings to individual meters, which would cost a tremendous amount, since each seperate meter would run about $200. But with some of the positive results we have seen from energy-saving campaigns in buildings that have master meters, it might make more sense to leave the meters as they are.

Some tenants of master-metered buildings in New York City have discovered that energy conservation can be an alternative to a rent increase. Residents of Silver Towers, a rental-apartment complex owned by New York University, were threatened with such an increase two years ago. Owner and tenants got together to discuss the problem. The owner agreed that if the electric bill could be reduced, the rent would stay the same. Some doubt was expressed that the tenants, mostly university professors, could cut down on electricity. And since the building was master-metered, there was no way to measure individual successes or to punish individual excesses. The tenants were not prepared to do anything inconvenient, such as drying their clothes on the roof or giving up hot baths. They agreed, however, to follow the usual minimal procedures: turn off unnecessary lights, keep drapes closed in summer, turn off air conditioners when nobody was at home. After a few weeks, there was evidence that the simple energy plan was succeeding. In fact, the tenants did better than to avoid a rent increase: they got a cash rebate from the owner. In 1974–75, before the conservation program, they collectively burned up 3.3 million KWH of electricity. In 1975–76 the energy plan was in effect, and they used 2.7 million KWH, or a savings of 660,000 KWH over the previous year. That's $66,000, at New York prices.

The Silver Towers results prove that renters can cut large chunks off utility bills without sacrificing comfort and without buying anything. Renters sometimes ignore the possibilities of conservation because they think they have to invest in storm windows or caulking, or other expensive items that will stay in the apartments when they leave. The people at Silver Towers learned that you don't need to go to the hardware store to ward off a rent increase. You can do it with light switches and thermostats.

Apartment dwellers can, therefore, follow somewhat the same kind of energy plan that this book proposes for homeowners. As for the energy manager, he does not have to be one single person; the job can be spread out among many people. And even though the building may have one big meter, it can be used effectively in the conservation campaign; somebody can read it and post each day's results—or a weekly summary of results—on a bulletin board. In that way, tenants can watch their collective progress.

The owners of the building might also agree, as at Silver Towers, to distribute the money that is saved, in the form of a rebate check. The rebate check offers an additional advantage over the individual homeowner's energy-saving program: the lone meter reader can calculate what he is saving, but there is no congratulatory check to certify the effort.

Experiments are now being conducted in various universities to test the value of such inducements as the rebate check. In one natural-gas study conducted by Johns Hopkins University, the results were not promising, but the researchers decided that one of the reasons for this was that the rebates were so small. Another experiment by the same university on reducing electric rates in apartment buildings seems to be succeeding. Renters are told (1) how to conserve, (2) sent various letters explaining the program, and then (3) provided with reports that describe what they have saved over predicted usage. The reports are accompanied by a check. In one building the energy savings went from 5.5 percent in the first two weeks to 9.9 percent in the second, and to 11 percent in the third. The checks got bigger, too. The experiment is not yet concluded, but early returns are promising.

What Co-Op Apartment Owners Can Do

The large co-op buildings, where people own their apartments and collectively maintain the whole structure, offer the greatest opportunities for conservation and profit. There are even enough ways to reduce utility costs to provide a full-time career for imaginative and/or unemployed residents, or starting an energy program could be a summer or a part-time job for one tenant-owner. This building energy manager could donate his or her services or take a percentage of the money saved—like a bounty hunter—for the plans that he or she devises.

One of the most persuasive illustrations of what a large co-op can do comes from Penn South, a group of fifteen twenty-one-story buildings housing middle-income families in Manhattan; twenty-eight hundred families live at Penn South. The co-op did not need to be told that its utility bills had increased from $780,000 in 1972 to $1,588,000 in 1975. The bill for oil had gone up 30 percent and the bill for electricity had jumped 100 percent. The co-op decided it would try an energy-saving plan, and David Smith, president of the tenants' organization, got the plan moving. It included the following provisions, which you may want to adopt for your co-op (or master-metered) building:

(1) Everybody replaced incandescent lights in kitchens and bathrooms with fluorescent lights.

(2) Some bulbs were removed from multiple-light fixtures.

(3) People agreed to turn off lights in unoccupied rooms.

(4) Air conditioners were used more conservatively, and never operated in empty rooms or apartments.

(5) Electric heaters were not allowed.

(6) People were encouraged to pool their elevator rides as much as possible.

(7) An "appliance fast" was held once a week, on Wednesdays. On that day, the use of such appliances as toasters, broilers, irons, dishwashers, and plug-in radios was forbidden. A dispensation was given for television—but for only one set, and preferably a black-and-white one.

(8) Heavy appliances were not used during peak hours.

In addition to these apartment commandments, the following modifications were made around other parts of the building:

(9) Incandescents in the lobbies and hallways were changed to fluorescents.

(10) Outdoor lights were put on photo-cell switches.

(11) Use of the laundry room was discouraged during the city's peak-load hours for electricity: 7 to 8 p.m.

(12) Some limitations were put on the amount of heat and air conditioning supplied to apartments. The central air-conditioning system was used at only three-quarter capacity.

The plan, according to all reports, has worked well. In two years the co-op has saved $410,000—which represents 700,000 gallons of oil and 2.7 million KWH of electricity.

The savings from such simple conservation measures can either be divided among the tenants, absorbed in a maintenance fund, or used for more sophisticated building modifications; co-ops may perhaps want to use the profits they get from turning off lights in order to finance a full-scale retrofit program that goes from the boiler room to the penthouse.

Residential buildings can learn from the energy-saving programs done by business buildings. A study performed on an office building at the Oak Ridge National Laboratories in Tennessee concluded that almost 45 percent of the energy used in that building—amounting to $16,500 a year—is saved without adverse effects or without excessive costs for building modifications. A Miami office-building manager raised indoor (air-conditioned) temperatures from 72° to 78°, allowed the humidity to rise from 50 percent to 60 percent, and is benefiting by $4,025 a year. A Minnesota office-building manager cut down on the amount of air that escapes from his building—another important option for residential buildings—and saved $2,750 a year. Two worthwhile corporate guidebooks are:

"Guidelines for Energy Saving in Existing Buildings"
Conservation Papers No. 20 and 21
Federal Energy Administration
Government Printing Office,
Washington, D.C. 20402
(Part 1, $5.25; Part 2, $5.05)

Total Energy Management Handbooks
National Electrical Contractors' Association
7315 Wisconsin Ave.
Washington, D.C. 20014

You can also obtain good material from your local utility companies, who often will offer free advice or even a free analysis of your building. Another good source is a local architects' association.

Energy-Auditing Your Building

Perhaps the most complete account of what you can do to tighten up a building is contained in "Guidelines for Saving Energy in Existing Buildings," Conservation Paper No. 20. It contains suggestions, charts, checklists, and case studies of buildings that have tried various modifications, and how much money they saved. The book also describes how to conduct a step-by-step energy audit of a building, the main points of which are as follows:

(1) Walk through the building, looking for lights that can be removed, extra appliances that can be discontinued, energy guzzlers that can be unplugged. This is conservation by reduction. If there are one hundred lights in your building that burn constantly, and they contain 75-watt bulbs, this is costing $9 a day at the 5-cent-per-KWH rate. That's $3,258 a year, which could be cut by one-third with 50-watt bulbs, and by much more than that by using fluorescents. Cutting on the lights also lightens the load on air conditioners in the summer.

(2) Find and replace broken windows, leaky faucets, dusty light bulbs; fill obvious cracks around door frames. Do the easy maintenance jobs. Even get a few things repaired by a carpenter or serviceman; it might cost a little money, but the returns can be as high as 15 percent off the current utility bill, according to the guidebook.

(3) Discover whether energy is being wasted in unoccupied areas; turn off lights in places where people never go; close heating ducts and unplug water heaters in unoccupied apartments.

(4) Check the water-heater temperature. You may be able to turn it down.

(5) Check the oil burner or furnace. A maintenance call can save thousands. A retail store in New York began getting its burner serviced for $500 annually, and reported a $3,600-per-year savings in fuel bills.

Attempting Larger Energy-Saving Projects for Your Building

You can't go too much further without expert help, but by this time you may have collected enough extra money to hire experts. Possibilities include: insulating the pipes in the basement; redesigning the furnace; installing a heat-recovery unit on the building's central air conditioner for free hot water; insulating apartment walls; putting up storm windows, awnings, or sun screens; installing thermostat timers in all apartments; and countless other possibilities. A New York office-building superintendent put spray nozzles on the lavatory faucets, reduced the hot-water temperature from 135° to 90°, and is saving $2,250 a year. What you choose to do depends on your building and its location and characteristics. You may need expert advice, and need to spend money.

Some tenants' associations are opting for even more ambitious projects, such as solar water heaters or space heaters, or even windmills. A sixty-five-year-old co-op on the West Side in New York is undertaking one of the largest solar-conversion projects in the country. But they never would have done it—according to one of the members—without a $110,000 grant from the Department of Housing and Urban Development. Energy-saving grants are more easily available to people in large buildings than they are to individual homeowners. If your building owner wants to shop around for a grant, have him check the sources in Section 11.

SECTION 11

The Public Goodies: Banks, Utilities, and the Government

Utility companies, banks, the government, and sometimes large corporations have all become cheerleaders in the cause of energy saving, cajoling consumers from the sidelines with advertisements, edicts, and patriotic appeals. While much of this encouragement may be heartfelt, some of it isn't; and some of the rest falls into the category of "This and a dime will buy you two kilowatts." When the bluster has subsided, the tangible and constructive help that large public or private institutions hold out to energy savers is harder to find. We have tried to sift through the rhetoric in order to discover for you the best places to get money, loans, grants, tax breaks, rewards, or useful help.

Getting the Most Out of a Utility Company

In the first place, these corporations look a little uncomfortable dressed up as energy-saving cheerleaders. As long as they have produced kilowatts, utility companies have advertised the advantages of electric- and other fuel-powered conveniences; they have encouraged and even subsidized developers to build more all-electric homes; they have even given out medallions to reward owners of such homes. Utility companies have also rewarded heavy industrial users of kilowatts with declining "block rates" that offer substantial discounts to large customers. Today the utilities face the unusual and ticklish dilemma in which they must ask their customers to use *less* of what they sell. They give out booklets and conservation advice. But many of them must hope that not too many people take the advice seriously.

These companies are confronted with dwindling oil supplies if they sell too much electricity, and dwindling profits if they sell too little. To survive, they have to come up with a good tightrope act on the power pole. Municipally owned and cooperative utilities have a better record of supporting real energy conservation than investor-owned utilities (according to the Environmental Action Foundation, a group that keeps track of such matters). The latter companies—which provide most of the power used in this country—want to befriend the energy conserver, yes, but without losing their cozy relationship with the investor. (A report on some of the good things that municipally owned utilities are doing to encourage conservation is given in Section 12.)

One way that electric utilities expect to be able to walk the tightrope is with off-peak pricing. This has already been tried in a few places, and chances are it will become common practice all over the country. Off-peak pricing is currently a voluntary program. Customers must agree to adopt the experimental rate structure, and in some cases pay for the new meter that must be installed to record the off-peak electric consumption. They will then be rewarded with lower rates for using their appliances at times when the citywide demand for electricity is small.

Electric companies favor off-peak rates because this probably won't affect the overall volume of kilowatt hours the company sells; it will just move things around a bit and level off the peak load. Utilities must provide generators to meet the high demand in the daytime hours; the rest of the time, the equipment is underutilized. Off-peak also reduces the need for new power plants, and provides better all-round use of existing equipment.

Off-peak operates on the same principle as the phone company's night-discount rates. Electricity used in the late-evening and early-morning hours will cost you a lot less than electricity used in the daytime. But switching to off-peak is a kind of gamble for the electric-utility customer. People who know how to work the system can save a lot of money; people who don't will end up paying much more for the same amount of electricity they now use. That's because the off-peak nighttime rates will be lower than the current standard rates, but the daytime rates—at least under the experimental system now—will be much higher than the standard rates. Unless you use *most* of your electricity at night, you lose with off-peak rates.

There are various formulas for off-peak, some of them more difficult to live with than others. The Con Ed utility in New York, in a current off-peak experiment with some households, defines the bonus hours as 10 p.m. to 10 a.m.; so, to get a lower rate, those people have to take their showers and do their laundry with Johnny Carson. The punishing aspect of the New York plan is that, among these households, those who continue to use most of their power in the daytime pay a high penalty. Consider a family that pays $26 a month for electricity in New York, or $312 a year. Under the off-peak system that family must use 70 percent of its electricity on the night shift in order to get a yearly discount of $32. But if that family only manages to use 50 percent of its electricity at night, it will *lose* money—and pay $35 *more* than the current rate. People who run most of their appliances in the daytime pay even more money than that. In the experiment, the on-peak summer rate jumps to 18.6 cents per KWH. The New York experiment is a case where the benefits of using the proposed off-peak rate are much smaller than the penalties for not using it.

Other electric-utility companies are experimenting with less severe off-peak price structures, and with more generous low-rate hours, such as 8 p.m. to 10 a.m. Even so, people who choose off-peak systems will have to do two things. They will (1) have to figure out how to run more of their appliances at night. And since most people turn down their electric space heaters and turn off the lights before going to bed, it will take imagination to find ways to use more kilowatts during evening hours; families can do the laundry, dishwashing, and shower taking after dinner, but that probably won't be enough to make off-peak profitable. And (2) they will also have to find ways to cut down on electricity consumption during the day.

One way of doing the latter is with some of the procedures described in Section 3, or with some of the devices presented in Section 4. Such procedures as putting the water heater on a schedule may be marginal for a normal utility-rate system, but they become very important under off-peak pricing; under such a schedule, you would turn off the heater during the day, when rates are highest, and heat all your water at night, when you get the discount. To get the most out of off-peak, you must also make sure that continuous support systems, such as the refrigerator *and* the hot-water heater, are set to require as little electricity as possible. Furthermore, insulating the hot-water tank so that it won't draw extra electricity to make up for heat losses becomes an even higher priority than it was before. Disconnecting the automatic defrost so that the refrigerator runs more economically also becomes a more profitable prospect. So does converting to fluorescent lighting and modifying the air conditioner for more efficient operation.

Check with your state public utilities commission about whether it is considering an off-peak rate schedule in your area. While you are at it, ask them how the commission regulates the electric rates. Some utility companies continue to reward the big users; others have adopted a more conservation-oriented rate. Whether your public utilities commission encourages or penalizes energy saving in its rate structure depends a lot on public pressure.

A few utility companies offer important advantages to energy conservers besides off-peak rates. A couple of the best ideas are described in the book *Taking Charge*, published by the Environmental Action Foundation and available for $3.50 from the foundation, at 724 Dupont Circle Bldg., Washington, D.C. 20036. They include:

(1) *Lifeline rates.* Yellow Springs, Ohio, was the first community to adopt a "lifeline": a very low basic

rate for the minimal amount of electricity needed to support human life. Lifeline rates are beneficial to poor people, and to older people who must survive on welfare and food stamps and cannot afford much electricity. Other utility companies are now adopting lifeline rates—some very reluctantly. Most, however, still reserve their highest rates for the smallest users. A mandatory "lifeline" rate is being considered as part of President Carter's energy plan.

(2) *Points of comparison.* Seattle's City Light includes with the monthly bill a comparison with how much was spent in the same period a year earlier. It is a built-in conservation yardstick, and a way people can keep up with their own energy-saving progress.

(3) *Insulation discounts.* The Kentucky Association of Electric Co-ops helps finance insulation for its members. The Warren Rural Electric Co-op Corporation in Bowling Green, Kentucky, bought a truckload of insulation and sold it to customers at cost.

(4) *A better meter.* PEPCO of Virginia is running a test to determine whether the installation of a Fitch energy monitor in the kitchen will encourage conservation. The Fitch monitor calculates, in dollars and cents, the amount of electricity being consumed in a house at any given time. PEPCO got a government grant to do the test.

(5) *Free advice.* Some utilities publish energy-conservation booklets, and a few will even come around to the house and offer on-site energy-saving suggestions.

There was some talk, in an early version of the President's energy plan, of forcing the utilities to finance the home-insulation programs, and then gradually to collect the money off their customers' utility bills. This idea was rejected in Congress, but the utility companies are increasingly aware that they have to do something to help people save energy. If you contact your own electric-power company, you may find that they have some sort of advice or financial support to offer. If they don't, ask them why not.

Banks and Other Loan Sources

There are several sources of money for financing your conservation improvements or solar installations. Probably the best idea is to borrow from a non-profit city corporation such as one planned for Hartford, Connecticut. Unfortunately, not many such programs are on the drawing board. But find out from your city government if any local money is available, non-profit, for individuals to retrofit their homes. A second possibility might be the utility company itself if yours has started a lending program. The state of Montana allows utility companies to lend money for energy-conservation and alternative-energy systems, at rates not to exceed 7 percent a year. The utility companies are given a tax credit to make up for whatever money they lose on this low interest rate. Various other utility companies around the country have loan programs. The government wanted to make utilities' lending a mandatory provision in the new energy law, but the idea ran into some stiff opposition in Congress and was abandoned.

Then, there are the regular banks and lending institutions.

A variety of financing plans might be available for solar installations, and which plan you choose will have a major effect on whether the solar conversion is worth the money. A booklet entitled "Home Mortgage Lending and Solar Energy," put out by the Department of Housing and Urban Development (Government Printing Office, Washington, D.C.; $1.40), explains the options. The booklet contains a chart showing how different kinds of financing affect solar-construction costs; it is reprinted as Table 13. These estimates refer to a solar space-heating installation, an expensive proposition. If you are putting the cost of solar into a first mortgage, you will probably have much smaller monthly payments than if you finance it through a second mortgage or home-improvement loan. Naturally, if you qualify for an FHA- or VA-

Table 13: How Financing Affects Solar Costs

ILLUSTRATIVE LOAN TERMS AND MONTHLY DEBT SERVICE UNDER PRIVATE-LENDER FINANCING ALTERNATIVES

LOAN TYPE

	First mortgage				Second mortgage	Home improvement	
	Conventional		Conventional	FHA	VA	Conventional	Title I	Conventional
Loan/Value Ratio	70%	80%	90%	93%	100%	75%	100%	100%
Interest Rate	8.5%	8.75%	9.0%	8.25%	9.0%	13.5%	11.5%	12.5%
Maturity (years)	27	27	27	30	30	10	12	5
Mortgage Insurance	—	.15%	.25%	.5%	—	—	.5%	—
Monthly Cost per $1,000 of Loan	$7.88	$8.16	$8.41	$8.06	$8.23	$15.23	$13.13	$22.50
DOWNPAYMENT FOR AN $8,000 SOLAR ENERGY SYSTEM	$2,400	$1,600	$800	$560	0	$2,000	0	0
MONTHLY COST FOR AN $8,000 SOLAR ENERGY SYSTEM	$44.15	$52.23	$60.53	$58.53	$64.37	$91.36	$105.07	$179.98

Source: Based on loan terms in the Boston area in March, 1976. Assumes that appraised value of system is same as full cost of $8,000. The chart comes from the HUD booklet "Home Mortgage Lending and Solar Energy."

secured loan, you have a much smaller downpayment, and probably longer to pay.

Whether various banks and lending institutions will finance your solar-space-heater, or insulation-retrofit, package depends on several factors discussed in the booklet. Some banks are still wary of solar installations for new houses, distrust the installation's potential resale value, and treat them as "overimprovements" or luxury items that are worth far less than their actual cost. Other banks are becoming more supportive of the idea of solar heating. The main thing to remember is that solar energy is a new concept for most people, and bankers react to it with the same diversity of opinion as does the general public. If one bank discourages a loan, the bank next door might be eager to offer one.

Another problem with the banks arises from how they calculate your ability to pay off a loan. Most of them follow the rule that housing costs should not exceed 25 percent of income, according to the booklet. A solar space-heating installation may increase the original cost of housing beyond what banks believe you can afford to pay. The hitch in the calculation is that, because of the solar heater, your energy costs will be reduced, thus making it that much easier to make the monthly loan payments. But banks do not usually consider the energy costs in their calculations. They use a formula called PITI—the principal of the loan, the interest payments, the taxes, and the insurance premium—in figuring your ability to pay. The HUD booklet suggests that banks add an "E" to PITI —representing the energy costs to maintain a home. If reduced energy costs are considered, more people would qualify for mortgage loans that include a solar installation. Ask your banker if he includes the energy factor in his loan formula.

Solar energy-saving improvements to old houses, the HUD booklet points out, are usually easier to finance than the same on new houses. When they finance a home improvement, bankers are interested mostly in the borrower's ability to pay, and not so much in the value of the property as collateral. Some banks are now advertising "energy-saving" loans at interest rates that run around 9 percent. Solar water heaters and even solar space heaters can be financed with such home-improvement loans.

Another option besides the home-improvement loan is to refinance the entire house to include the energy-saving additions. Many banks will be eager to refinance your home, because they can then raise the interest rate. Whether you would benefit from such a move—as opposed to a regular home-improvement loan—depends on comparative rates of interest, the amount of downpayment necessary, and the cost of the improvement(s) you wish to undertake.

Federal monies will be available, as proposed by the Carter Energy plan to open up the energy-saving loan market. Things may change in the next few months: rates may be reduced for energy-saving loans, and local money may be easier to get. If you pay close attention to what is offered by your local banks and lending institutions over the coming months, you may be able to finance a project that was previously out of reach.

There are also the usual FHA and VA programs that can be applied to solar or insulation expenditures. Only a small percentage of people can qualify for these guarantees, but the attitude of FHA does influence other lending institutions. Solar retrofits can be effected through the FHA Title One property-improvement program; in fact, Title One was amended in 1974 to include solar expenditures in the loan insurance. HUD field offices have been issued guidelines on how to handle mortgage-insurance requests for solar devices, and FHA has incorporated standards for solar heaters into its Minimum Property Standards. FHA field offices now insure for solar hot-water heaters, according to the HUD booklet. Check with your own bank, credit agency, or local FHA office to see whether these benefits will apply to you.

Help from the State and Federal Governments: Energy Saving and Taxes

You might have read it in *Time* magazine. William Nagy, Long Island homeowner, was encouraged by the President's energy address, and by his own soaring fuel costs, to go out and buy a solar water heater. He made some careful calculations, and figured that he could save $250 a year on fuel bills. Since the unit cost $1,600, the break-even point would be eight years —not the kind of return that Wall Street likes, but one that made economic sense to him, and one that also had its patriotic aspect. Nagy expected that the outer reaches of government would support his solar conversion as much as the President would. But, the next thing he knew, he was advised that the property assessor was adding $90 to his yearly property tax if the installation were to be made. It was enough to throw off all Nagy's calculations—as well as his assumption that if the homeowner does his part, so will the government (in this case, state). Nagy canceled his solar contract.

Some states have recently passed laws that encourage energy conservation and make things like solar water heaters more economically attractive. Other states are still working under laws that inhibit and discourage such projects. There isn't room here to report on all the various regulations, on a state-by-state basis, so the best thing for you to do is check with your state energy office. Before you make any major improvements—adding insulation, a solar water heater, a solar space-heater, etc.—check on building codes, property taxes, solar easements to protect the sunlight that falls on your collector, and whatever economic benefits might be attached to your project. The extra incentive, or *dis*incentive, provided by your state government may be a deciding factor in whether or not you should pursue your energy-saving idea.

Several states have added their own incentives for people who go solar. You'll have to check with your state energy office to get the specifics, but here are some of the more interesting provisions:

Arizona allows people to write off the cost of solar installations from their state income tax. It can be amortized over a three-year period. Arizona also gives a total exemption from the property tax that would be charged for the added value of the solar unit. Several other states do the same.

California allows a state-income-tax credit of 10 percent of the cost of solar installations. The total credit cannot exceed $1,000.

Colorado assesses the value of all solar systems at 5 percent of their original purchase price, and thus effectively limits the property taxes attached to solar units.

Georgia refunds the sales tax paid on solar equipment. *Texas* also exempts solar products from sales taxes.

Maine excuses solar equipment from property taxes for a period of five years. That state has undertaken a large-scale insulation program for low-income families. Maine has also mandated a utility-rate structure that encourages energy conservation.

Massachusetts exempts solar heaters and windmills from real-estate taxes. The state requires that estimates be taken on solar collectors and wind generators for every new public building constructed in the state. Equipment installed in all such buildings must be chosen on the basis of a life-cycle cost analysis. Massachusetts also has a corporate tax incentive for the use of alternate energy sources.

The National Solar Heating and Cooling Information Center keeps up with the changing state legislation on solar energy. You can write for a list of the most recent state laws. Address your request to the center at P.O. Box 1607, Rockville, Maryland 20850.

Federal Benefits

This year, 1978, the federal government is planning to offer some major tax breaks for people who insulate their homes or who add on solar water or space heaters.

If you are retrofitting your house for energy conservation, under the proposed Carter energy plan you could deduct 25 percent of the first $800, and 15 percent of the next $1,400, you spend. That's a maximum saving of $410 off your taxes. You can get the deduction for any energy project that is on the approved list published by the government. Most reasonable projects will be on the list.

If you put in a solar water or space heater, the Carter plan would allow even more of a tax advantage: 40 percent of the first $1,000, and 25 percent of the next $6,400, that you spend for qualifying solar equipment. That amounts to a total $2,000 you could deduct from your taxes for a $7,400 solar heater. You would have to buy soon, though, because these large deductions apply only to the taxable years 1977, 1978, and 1979. During the period of 1980–1981, the deductions are reduced to 30 percent of the first $1,000 and 20 percent of the rest of the money you spend on solar equipment, for a maximum deduction of $1,580. In 1982–83–84, you get only a 25-percent deduction of the first $1,000 and 15 percent of the rest, for a maximum tax break of $1,210.

Such tax advantages can contribute a lot to the economic attractiveness of solar heaters, and the tax savings should be included in your estimate of whether your solar heater or insulation will pay off.

The Solar Hot Line

The federal government now runs a telephone hot line to answer questions you may have about solar energy, or to provide technical information if you are building or buying a solar unit. The number to call at the National Solar Heating and Cooling Information Center in Rockville, Maryland, is 800-523-2929. In Pennsylvania, it is 800-462-4983.

Energy-Extension Service

ERDA has also sponsored an energy-extension program, similar to the well-established agricultural-extension program, to advise people in various localities about how to save energy. The pilot program has begun in ten states, and will give information and personalized assistance to individual citizens, businesses, schools, and government agencies. Since this may be the kind of specific, localized advice you have been looking for, you could benefit by contacting the energy office in the states that have started the extension service. Those states are: Alabama, Connecticut, Michigan, New Mexico, Pennsylvania, Tennessee, Texas, Washington, Wisconsin, and Wyoming. If the pilot program is a success, the government may finance energy-extension agents on a nationwide basis.

Grants

Big Grants

Money is available for solar-demonstration projects for housing developments, cooperatives, and big buildings. To find out what is available, contact the grant office at the new Department of Energy, the Energy Research and Development Administration (ERDA), or the Department of Housing and Urban Development (HUD). Getting large grants from the government is not an easy process, but money is available for energy projects, and *somebody* has to receive it. Your own county grant coordinator will probably have information on the types of grants available. If he or she doesn't know, he should learn.

Little Grants for Backyard Inventors

Popular Science gives $200 for the alternate-energy idea of the month. Its program called "Adventures in Alternate Energy" asks that readers, tinkerers, and local inventors who have turned their ingenuity to solving energy problems in low-cost, practical ways submit their ideas for possible publication. The magazine describes and illustrates each prize-winning idea in a regular feature. It has awarded money for such inventions as: the solar window; various kinds of do-it-yourself solar collectors; the flue radiator, which provides free hot water from heat that goes up the chimney; etc. If you have conjured up a practical energy saver, *Popular Science* is a magazine which can get your idea some public attention. Write to Energy Adventure, Popular Science, 380 Madison Ave., New York, N.Y. 10017.

The National Bureau of Standards, a government agency, also has a grant program for small inventors. You have to send a written description of your invention to NBS; and if it qualifies, the idea will be submitted for a technical review. If it passes that hurdle, there will be a second, more thorough evaluation. If NBS still likes the idea, they will forward it to a federal energy agency, and you as inventor will be eligible for a direct grant. It's a long process, but if the idea survives, the government could give substantial financial assistance to further its development.

To submit an idea, or to get more information, write to the Office of Energy-Related Inventions, National Bureau of Standards, Washington, D.C. 20234.

SECTION 12

"Turning Down" the Town: They Did It in Seattle

It's hard to make the connection between how *you* run your appliances and how *they* run the power plants, or to believe that your little roll of insulation will tip the balance in the national debate over nuclear power, offshore oil, and other matters of high energy policy. What happens on your side of the meter seems somehow light-years away from the utility companies, oil companies, and coal magnates that decide how to feed it. But the physical connection—paradoxically—really *is* direct: you are tied into the utility company with real wires.

The absence of any real conviction that citizens' personal habits of energy economy have much to do with utility policy was shown in the recent New York power failure. Nowhere in the pages of major newspapers, through article after article about the social and political consequences of the blackout, was there even a small mention of energy conservation as a way to avoid such occurrences in the future. High electricity demand of course did not *cause* the blackout, but it certainly contributed to the problem. The blackout came during peak summer hours, when the entire town was running on High Cool and the utility company could not shed enough of the load to maintain power. In the vast postmortem, there were discussions of new power plants, of the utility company's ineptitude, of the city's reaction, of how to pay for the spoiled meat and melted ice cream; but energy saving did not come into it. The federal government continues to assure people that their efforts to save energy are somehow vaguely connected to the national interests, but conservation is generally ignored as a specific solution to a local problem. New Yorkers and their newspapers were replete with criticism for Con Ed's inability to deliver power. Con Ed may not be the best utility company in the country, but nobody pointed out that if its customers could reduce their consumption, Con Ed's performance might automatically be improved. The company would be better able to meet peak demands, and might even require fewer new generating plants. An ambitious, citywide conservation program, begun as a response to a blackout, would give consumers something better to do than fatalistically complain. They could help shape their own energy destiny.

If this sounds like a Pollyannish idea, it isn't. It did happen in Seattle, Washington, two years ago.

The Successes of Seattle and Los Angeles

Seattle's local utility, City Light and Power, one of the more innovative electric companies in the nation, was considering whether to build a nuclear power plant. Nuclear power is an issue that has divided communities all over America, and expert after expert has become engaged in the debate over safety and inevitability. Individual utility customers usually don't have much say in the matter. But the Seattle City Council, governing body of this municipally owned utility company, decided to pay more than lip service to public participation. What happened, according to a report in the newsletter *Power Line,* published by the Environmental Action Foundation, ". . . was an extraordinary exercise in cooperation between utility officials and the public. Economists, sociologists, engineers, planners, transportation experts, historians, recreation authorities, and community leaders were brought in to prepare what the utility manager called 'the most thorough energy analysis ever undertaken in this country.'" The study, entitled "Energy, 1990," analyzed alternative methods of meeting Seattle's energy needs over the next fifteen years.

The alternatives included some unusual items like solar electric generation, but the most unusual of all was conservation. The study said that a certain amount of energy saving was feasible—feasible enough to be a utility strategy, a kind of invisible investment, like building a new plant. And when conservation becomes utility strategy, each individual electricity consumer becomes a policymaker, knowing that his or her decisions on where to set thermostats will have a direct bearing on the success or failure of the utility company's plan. In Seattle, energy saving became more than an individual's money-saving opportunity, or some vague expression of patriotism. People were *voting,* in a sense, with their meters.

Among all the options, the city chose the one that called for the most conservation. An energy-saving goal, much like the one you may have set for your home, was devised for the entire city. Everyone would have to do his part, or else the utility company would be hard-pressed to meet future demands. Seattle residents did not have to give up personal comfort in order to give up the nuclear power plant, because the study also concluded that a high standard of living could be maintained even while using much less energy. The problem was how to motivate consumers to conserve.

Seattle City Light responded with an ambitious program that makes the customary utility-company practice of stuffing a few energy reminders into envelopes and paying for a few conservation-oriented advertisements look meager indeed! The company did all the usual public-relations things, but it also offered a Home Energy Loss Prevention program. A utility-company truck is sent out to individual houses, where the company will install insulation jackets on hot-water tanks for $9.50, or about half the price of a store-bought retrofit kit. During the home visit, utility technicians will also advise on how to ventilate the attic, how to seal up the sides of the fireplace, how to insulate, caulk, and weather-strip. Seattle City Light trucks also carry loads of low-wattage light bulbs, which are sold at cost.

The city, with the support of the utility company, has devised an unusual heat-loss standard for new electrical hookups. If the new building, or the building converting to electric power, can't meet the insulation and other heat-loss requirements, electricity will not be provided.

And finally, the utility company has taken to the air, using a fly-over scanner, an infrared device that takes pictures of heat losses on the roofs of Seattle homes and other buildings. (See Plate 14 for an example of a portable scanner.) The pictures help show people the need for more insulation.

For all these conservation activities, Seattle City Light and Power spends over $350,000 a year—to convince people to use *less* of its product. This may sound highly unprofitable to a privately owned utility company that must satisfy investors, but for a municipally owned company it makes good sense. All the promotion has resulted in a 10-percent drop in electricity demand over the last few months, and the reduction has already saved the company much more than the cost of the conservation program. A severe drought in the Pacific Northwest has forced Seattle City Light and other utility companies to buy power from other sources, and at higher prices than the power they usually produce in their hydroelectric generators. But, according to a utility-company spokesman, conservation saved City Light from having to purchase eleven million dollars' worth of high-priced electricity. And the city has now reached one-third of its 1990 goal for energy conservation—which puts Seattle far ahead of schedule.

Many other utility companies likely wouldn't dare to use energy conservation as an assumption in deciding on their future company investments. "How can we be certain that people will conserve?" they would ask. Yet there is evidence from places other than Seattle that people *will* respond, if they are given a goal to reach and a specific reason for doing so.

The City of Los Angeles discovered how much power could be conserved almost instantaneously, without any advance preparation at all, during the energy shortage of 1973. The city was low on oil and had to institute emergency energy saving. The city notified merchants that they would have to cut energy use by 20 percent; and homes and industries were advised they would have to cut by 10 percent. Within four days—*before* people had a chance to install insulation or buy gadgets!—the city was running on 11-percent less power; and by the end of the first week, electric consumption was down by 14 percent. During the first two months of the crisis, Los Angeles used 17 percent less electricity than it used during a corresponding period a year earlier. The Los Angeles emergency plan, viewed as a major success, was studied by the Rand Corporation.

Rand searched for reasons to explain the unprecedented citizen response.

One reason, Rand decided, was that the Los Angeles plan had teeth. There were stiff penalties for people who exceeded the power quotas: for commercial customers, the first excessive utility bill would carry a 50-percent surcharge for the amount of the entire bill; the second would result in complete power cutoff for two days; and the third, a power cutoff for five days. Commercial customers also faced an absolute ban on decorative outside lights, and restrictions on most neon signs and movie displays. Billboards could be illuminated for only a few hours a night; heating and air-conditioning units in business establishments had to be completely shut down during non-business hours; and the thermostats in those businesses had to be set no lower than 78° for cooling and no higher than 68° for heating.

According to the Rand study, the teeth worked. The commercial sector, which uses 50 percent of the electricity in Los Angeles, achieved a 38-percent reduction; industry, which uses 20 to 25 percent of the total power, made a 10-percent reduction; and residential users cut back by 18 percent. A more voluntary approach might have worked, too, but Los Angeles did not have the time to develop such a program. The amazing thing is that people did not complain about the sanctions. They found they could achieve most of the required reductions in light bulbs alone, and could make the 10- to 20-percent drop in usage without hardship. This, indeed, is additional proof that about 20 percent of the average electric utility bill *can* be removed through quick, free, and painless surgery.

In fact, conservation was so easy that once people got into the habit, they continued to use less electricity long after the sanctions were lifted. Even into the summer of 1974, which was hotter than the previous summer, Los Angeles operated on 14-percent less electricity than it had used the year before. The people at the Department of Power and Water, the city utility, tell us that conservation continues, even today. Elizabeth Wimmer, manager of media relations, says, "People never go back. Once they get into the ways of conservation, once they get along with fewer lights or less water, they keep it up forever." The Los Angeles crisis has long been forgotten, but the city's overall electric demand is still 4 percent lower than it was in 1974 —and even though the city has grown since that time.

Should Energy Conservation Be Made Mandatory?

The success of the Los Angeles program raises a question: Should mandatory conservation wait for a crisis, or should it be part of national energy policy? The federal government, so far, is committed only to *asking* people to save energy.

The government tested voluntary programs in two states, Massachusetts and New Mexico. The idea was to put on a media energy blitz: newspapers wrote about conservation, politicians talked about it, various displays circled around shopping centers, television stations put on energy specials. The most expensive part of the Massachusetts program was the energy questionnaire, which families filled out and sent to Washington. They got back a complete energy profile of their homes: how much insulation they needed, how much it would cost, how much fuel and money would be saved... The most imaginative part of the program was the Red Rover, a truck with an infrared scanner. The scanner took thermal pictures of houses, to show hot spots—places where heat is leaking out during the winter. The results from the scanner were used to show homeowners how to reduce heat losses. According to Charles Bowser, who supervised the Red Rover operation, most of the findings were predictable, but still useful. People responded to the idea of insulation when they could see a picture of heat escaping from their houses. In some instances, the Red Rover could also pinpoint unexpected sources of heat loss, which showed up around fireplaces and chimneys.

Enough hoopla was created around the Massachusetts energy program that 140,000 people, or 15 percent of all the homeowners, completed the questionnaire, and got home energy profiles from the Federal Energy Administration. Since the questionnaires were confidential, there is no way to discover how many people actually later refitted their houses, or how much fuel each of them saved. During the one-year campaign, the state also offered a toll-free energy hot line, adult-education classes in conservation, and even a couple of insulation cooperatives, where people in groups could get a better price for the material than they could have individually. And the state did not stop with the one-year program. It has developed an ongoing energy program through the Massachusetts energy office. The state has already passed an appliance-labeling law that is more stringent than its federal counterpart; refrigerators already carry an energy label. It is adopting new building codes, with higher requirements for insulation and with other energy-saving provisions. It has passed laws offering incentives to people who install solar hot-water or space heaters. But the success of the Massachusetts program rides or falls on public enthusiasm, which takes a lot of money to arouse and a lot more money to sustain. There was no specific goal in Massachusetts, as there was in Seattle and in Los Angeles. The one-year media blitz cost the state and the Federal Energy Administration approximately $500,000.

It is not clear whether the government will continue such voluntary efforts, or will expand them to other states. The results from Massachusetts were promising—an 18-percent drop in fuel consumption during a very cold winter—but nobody knows how much money it would take to reach the rest of the Massachusetts population, or to encourage the original supporters to continue their efforts. Plenty of federal energy money will be flowing into the states in the next few years, but each state will find its own ways to spend it. Some states will offer energy-extension services, which will work like the agricultural-extension program. Others are planning adult-education courses, solar-demonstration projects, and the like. Some may offer the services of an infrared scanner, like the Red Rover. Some have passed legislation that requires better appliance labeling or more energy-conscious building codes. A phone call or letter to your state energy office, or to a local state representative, should get you up-to-date on the energy services that your state will provide.

The Hartford Plan and Local Energy Conservation

State energy programs can be helpful, but they are often too diffuse, and too far removed from the people, to generate the kind of enthusiasm we saw in Seattle and Los Angeles. Most of the best energy-saving ideas continue to be developed at the local, or city, level.

The city of Hartford, Connecticut, plans to go even farther than Seattle. The municipality itself will train people to conduct energy audits in public buildings and private homes, through a non-profit energy corporation. The corporation will create a lot of jobs by hiring local people and teaching them to be energy technicians. The technicians will visit homes and businesses, advising people on the best retrofit package for their specific situation. They will offer information on the full range of possibilities: insulation, storm windows, furnace modification, solar heating, etc. The hope is that people will trust the energy technicians more than they trust insulation salesmen—who are promoting only one product.

The details haven't been worked out yet, but Hartford also hopes to solve one of the thorniest problems of energy conservation: How to finance it. Many people like the idea of insulation, but not as much as they dislike the 9-percent bank loan to buy it. The Hartford energy corporation wants to work out an arrangement with the regional water and sewer authority, an organization that can levy liens on personal property, and can therefore offer very low interest rates on loans. The corporation would eventually like to create a kind of municipal energy bank, where the property owner could get a loan with no downpayment. The monthly payments on the loan would be carried in the utility bills. That way, a homeowner could actually finance his conservation through his own savings on energy. The cost of insulation would be directly offset by the lower utility bills that the insulation helped produce.

The Hartford Plan is one of the most ambitious in the country. Even if the financing idea falls through, the program answers one of the most frequent criticisms leveled at energy conservation: that conservation somehow threatens jobs. In Hartford, jobs will be created, not lost, through the corps of technicians.

We read about the Hartford Plan in a book called *New Directions;* the book describes a lot of innovative thinking being done by state and local governments. (If you want a copy, the address is at the end of this section.) *New Directions* contains an entire chapter on energy; and the article on the best local-conservation plans comes from the Center for Science in the Public Interest. The center reports the following:

The Burbank, California, utility, along with the city, devised an ordinance that restricts decorative lighting and other superfluous uses of electricity. The city achieved a 17-percent reduction in energy demand.

Los Angeles operates a solar-heated municipal swimming pool.

Santa Barbara, California, uses college students to canvass for conservation.

Pitkin County, Colorado, has produced a land-use code which includes energy-conservation requirements for new developments.

Danbury, Connecticut, has a civic-preparedness plan for energy emergencies.

Jacksonville, Florida, publishes home energy-audit packets.

Ocala, Florida, installs solar water heaters and then lets homeowners lease them from the city.

Helena, Montana, lets the developers finance the houses, but the city finances the solar collectors.

Williamsport, Pennsylvania, winterized homes for low-income families as part of a state-supported program.

The Cranston, Rhode Island, community action agency created a small business to build solar water heaters for poor people.

Vermont provides for an energy coordinator appointed by each town, to set up programs, look for money, and offer help to local residents.

Local energy projects like the ones described above have a habit of getting bigger, of expanding into new territory. The entire town of Davis, California, became energy-conscious as a result of a simple building survey done a few years ago. As we mentioned earlier, the survey showed that identical apartments could achieve much different indoor temperatures, depending on which direction they faced. An apartment oriented in one direction would require more air conditioning, and therefore more electricity, than an apartment oriented in a different direction, even if both apartments were part of the same building. This conclusion convinced the town planners to revise the building codes, and, after that, the town began to alter the entire system of land-use planning in order to encourage conservation. Sun-rights laws have been passed, protecting people's access to sunshine for solar heating. Builders are encouraged—and in some cases required—to face buildings in certain directions to maximize the benefits of sun, shade, and wind. What started as a small research project ended up giving Davis a new identity. The town of 35,000 people, most of them attached to the university, is known as the most conservation-minded community in America.

Energy saving has also given Davis a focus for community spirit. The entire town is behind the changes in local laws that again permit the use of clotheslines; that now allow fences to be put farther away from houses, where they don't block off winter sun; that mandate narrower roads which use less asphalt and don't collect as much heat. The city itself has built model solar homes, and has conducted detailed residential energy surveys to help people figure out where to cut energy use. The results have been a zero growth in electric demand per customer, and a 10-percent reduction of electricity use in the peak summer months.

Unplugging Your Town

You might encourage *your* community to name an energy coordinator and create some of the local programs listed above. Or, ask your local government the following questions:

(1) Does the town have an emergency plan, to provide for possible blackouts or fuel shortages? Such predicaments are less remote than natural disasters, for which many communities are at least partly prepared.

(2) Is there federal or state money available for energy programs in your locality? Are there winterization programs for people who can't afford insulation?

(3) Could the town rent an infrared scanner and do a heat-loss analysis of all the houses in the area, telling people the best places to put their insulation?

(4) Could the town buy insulation in bulk and sell it at cost to local residents in order to save them money?

(5) Is it possible to form a local energy corporation, similar to the one in Hartford, Connecticut?

(6) Could prizes be awarded for the best energy-saving ideas, or for the families who cut down their

electricity consumption the most over a fixed period of time? Could a local energy-saving campaign, like the one in Massachusetts, be established?

(7) Could energy be saved in public buildings, such as the city hall? Could the city pass public-building-code amendments or lighting requirements that reduce consumption of electricity?

Some localities doubt the ability of utility companies to continue to provide them with power. An unplugging process is beginning, reversing the entire trend toward rural electrification that began in the Roosevelt Administration. Furthermore, people look to depleting oil supplies, uncertain nuclear power, and a messy coal industry, and conclude, too, that their town might be better off producing its own energy. Some towns even believe they can make and distribute energy *more cheaply* than the utility company that now serves them. A wide variety of innovations are available. They include: producing methane gas from a city garbage dump; using heat from under the ground to keep houses warm; recycling waste oil; trapping the exhaust heat from power plants; and large-scale solar generation. It would become too complicated to describe all these options in detail. The important point is that many of them work right now, and some of them are economically less burdensome than the utility costs that cities now pay.

The Center for Science in the Public Interest, in Washington, D.C., has also kept up with these developments, and cites the following: Cuttyhunk Island, Massachusetts, will be getting most of its power from a windmill what has already been constructed. A few small cities in California have formed a geothermal consortium, to siphon heat from their geysers. Los Angeles already gets methane gas from its sanitary landfill. Ames, Iowa, burns solid-sewage waste for electricity. Some cities recover waste oil from automobiles and industrial plants, and use that oil again. Bridgeport, Texas, tiring of high electric rates, has decided to build a solar electric plant.

The prospect of energy independence is not limited to political entities like cities or towns. Even some large housing projects are experimenting with a program—called MIUS, or Modular Integrated Utility System. The idea behind MIUS is that large groups of people can economically generate their own power; use the waste heat normally given off by a generating plant to heat homes and apartments; and recycle water, sewage, and garbage to minimize their dependence on outside sources. MIUS is one possible answer for developers who are not allowed to build in certain cities because water and power resources are too limited. The Department of Housing and Urban Development has financed a large MIUS project for a group of residential buildings in Jersey City, New Jersey. Similar projects are being considered in other parts of the country.

Unplugging an entire town or housing development won't happen overnight. It takes planning, vision, and money before people can achieve complete energy independence. But as the problems of utility companies and fuel-oil suppliers grow more complex, energy independence is going to look more and more attractive. Some towns are making elaborate preparations for the future. The University of Minnesota has drafted a plan to make the town of Winona completely self-sufficient for energy by the year 2000. The plan provides for solar collectors; it also requires local industry to rely more on human labor and less on machine power for its future production. If you are interested in the details, you can write the Energy Design Studio, University of Minnesota, 2818 Coco Ave. SE, Minneapolis, Minnesota 55414.

For Further Information . . .

You can get more information on how local communities are solving their energy problems from the following sources:

Taking Charge: A New Look at Public Power, published by the Environmental Action Foundation, 724 Dupont Circle Bldg., Washington, D.C. 20036. The book, which describes some things the better utility companies are doing, and ways to make the worst ones do the same, costs $3.50.

Center for Science in the Public Interest now called The Citizens Energy Project, 1518 R Street NW, Washington, D.C. 20009. This group keeps up with local energy-saving activities, among other things. They provided the list of what various cities and towns are doing to conserve energy.

New Directions, published by the Conference on Alternative State and Local Public Policies, 1901 Q St. NW, Washington, D.C. 20009. The book is full of information about constructive things that communities are doing today.

SECTION 13

Is There an Energy-Saving House in Your Future?

Some of the more progressive architects and builders are now producing homes that are two to three times as efficient as the old fuel-waster homes, and at about the same cost. Other builders continue to design, make, and sell fuel wasters. A prospective buyer or buyer-builder who doesn't consider the energy-using factor in his calculations, as a kind of homeowner's life-cycle costing, could be in for an unpleasant surprise. The house down the street from his new one, which was purchased for a comparable price, may stay warm on one-third the fuel bill.

Enough information is available so that the careful home buyer can have a clear idea of what kinds of houses save energy. We can summarize some of their major features. We will get into these details in a moment, but even more important than these is the *change in philosophy* that energy-conscious home buying requires...

The innovative architecture of the recent past tried to capture the beauties of nature; and all those glassed-in specials brought the outdoors right into the living room. The irony was that most of these "natural" houses ignored the effects of Nature completely. If the rolling hillside was on the cold, northern exposure, then that was the place for the picture window. It didn't matter that the pleasant view had to be supported by more oil and a bigger furnace in the basement. Energy-efficient architecture, on the other hand, in all its various forms shares one characteristic: These fuel-miser houses are less *spectators* of nature and more *users* of nature. They employ the sun and wind as allies in producing comfort, which is after all the original purpose of home shelter. Even the most gadget-oriented of energy-saving homes try to rely on the sun, wind, earth, and trees as *natural heaters* or *air conditioners* first—and as *sources of visual pleasure* second.

The Old-Timer Factor: Traditional Local Knowledge

When you buy or build a house, the first thing to consider, in a general way, is whether the structure (or plan for the structure) and the overall "look" of the house was created to make a pretty picure or to conform to climatic realities. One way to decide is to compare the house to some of the older ones in the area; in other words, to consider the "old-timer factor."

It doesn't take a Ph.D. in thermal physics to discover that you can dry your hair in the sunshine, and there is something reassuring in the fact that a lot of the computerized, complicated energy studies take us right back to some simple and traditional building procedures.

Wherever you happen to live, there are old houses, or people who remember old houses—houses that managed to provide livable conditions before all this easy energy was available. We're not thinking about the fancy, drafty Victorian mansions that depended on ten fireplaces and an equal number of servants to chop the wood; we're thinking about simple shelters built by ordinary people. Their solutions to the problems of cooling and heating relied on local resources and on a knowledge of local weather conditions; and some of those solutions are still valid today.

Our first advice for anybody who is planning to design a house (or, in fact, to buy a house that has already been designed and/or built) is to check with local old-timers to see how *they* used to do it. In the southern part of Florida, a traditional design exists that offers plenty of indoor comfort without air conditioning. It is a house on stilts, with porches stretching around all sides, a large overhang, and a cupola in the middle where hot air can escape through a kind of structural attic fan. When air conditioning got cheap and available, people forgot about these early designs, but the most progressive architects are going back to them again. One of the authors of this book has such a house, cooled with the overhang and lots of roof insulation, where the air circulates with old *Casablanca*-style attic fans. The neighbors, who pay up to $150 a month to keep cool with air conditioners, are often surprised at how comfortable this stilt house actually is, and they ask where the idea came from. The idea, in fact, came from early Florida settlers, ancestors of some of the very people who ask the question. The information had somehow been blotted out in our less than three decades of air conditioners.

This type of knowledge is available in most every area of the country. In the Southwest, energy-saving houses are again relying on adobe brick, and even underground rooms. In the colder climates, architects are discovering that the best insulation is a pile of earth pushed up against the north side of a house. People who live in the cold farm areas of the North Carolina or Tennessee mountains scratched their heads in wonderment when urban dropouts brought high cathedral ceilings to the country with them; the locals had known for centuries that low ceilings keep people warm.

There is usually a reason for the variations in local home designs—a reason that goes beyond the expression of an owner's personality to an understanding of how to survive the weather—and be comfortable—on the area's resources.

Site Selection and House Orientation

In the past couple of years, researchers have spent thousands of dollars to discover at least one thing that the old-timers could have told them: *where* you build a house can be as important to using less energy as *how* you build it. Site selection determines, to a great extent, how hard your furnace or air conditioner will have to work, and how much you will have to pay for fuel. And by site selection we don't mean Florida vs.

Maine; we mean one part of your building lot vs. another part. If you are skeptical that such a choice could have any real economic repercussions, consider some of the findings published in a book, *Window Design Strategies to Conserve Energy,* which we obtained from the National Bureau of Standards:

(1) An experiment was conducted by the Lake State Forest Experimental Station in Nebraska on two identical houses. One was fully exposed to the wind and the other was protected by some shrubbery. The exact fuel consumption for maintaining an indoor temperature of 70° F. for each house was measured. A savings of 23 percent was recorded for the protected house.

(2) A fully exposed electrically heated house in South Dakota required 443 KWH to maintain an inside temperature of 70° from January 17 to February 17; an identical house sheltered by a windbreak required only 270 KWH. The difference in average energy requirements for the whole winter was 33.92 percent.

(3) George Mattingly and Eugene Peters of Princeton University are studying the effects of wind on a group of townhouses at Twin Rivers, New Jersey. Results, from scale models in a wind tunnel, suggest that a five-foot-high wooden fence would reduce air infiltration 26 percent to 30 percent. A single row of evergreen trees as tall as the houses would reduce air infiltration 40 percent; and the combination of the two would reduce air infiltration by 60 percent.

Proper site selection offers such inexpensive and easy energy savings that even people who install and use fuel-conserving equipment start out by carefully siting their homes. Evan Powell is such a person. He lives in South Carolina, and writes on energy topics for *Popular Science* magazine. His house is one of the most active energy laboratories in the country; Powell has installed a specially designed furnace, water-saving toilets and showerheads, etc. His fuel bills are much lower than those of his neighbors.

To begin with, Powell put Nature in his corner. He built the house on the south side of a hill, and the hill itself became a perfect natural windbreak from the north wind. The glass in the house mostly faces southeast, where the sunlight can get in and help out his furnace in the winter. Without this planning, his fuel bills certainly would not be as low as they are.

Every building lot has different characteristics, so we can't range too far in explaining where to place your house in order to save energy. The Owens-Corning people put out a booklet called "44 Ways to Build Energy Conservation into the Home." We found it to be one of the most uncomplicated, readable, and useful publications for people who are designing or buying a home. You can get the booklet from Owens-Corning Fiberglass Corp., Insulation Operating Divn., Fiberglass Tower, Toledo, Ohio 43659. Here are two of the house-locating principles that Owens-Corning espouses:

(1) In sunny climates, it is thermally advantageous to have the ridge of the house about parallel to the east-west axis. In cold climates, a north-south ridge would be best.

(2) In cold climates, orient the house south if possible, and, if appropriate, use proportionally more glass on the south wall, shading it with sufficient overhang to reduce heat gain in the summer. In hot, arid climates, consider a design or orientation that will buffer against hot breezes. Occasionally it is possible to locate the dwelling, or windows, to take advantage of the shadows caused by existing trees in order to reduce solar heat gain in the summer. Locating the air-conditioner compressor where it will be shaded—particularly in the afternoon—by the house, trees, garage, or carport will increase its efficiency and reduce energy use.

Such considerations may affect the piece of property you choose to buy (or whether you buy an already-constructed house that is poorly sited and/or oriented). Many subdivisions were laid out with more thought to traffic patterns than to weather patterns, so that when all the houses face the street, some may face precisely the wrong way for energy saving. You will save a lot of money by finding a development, or a lot within a development (as with, even, an apartment in a building), that gives you the best natural headstart on the heating or cooling job. The best energy-saving property is also the least expensive. In areas with cold winters, the best place to build is often on the south side of a hill, like Powell did.

Design of the Energy-Saving House

Somehow, energy-saving houses have been associated with certain bizarre architectural features that don't exactly blend in with a regular suburban neighborhood. Cave houses, underground kivas, solar huts, ice-block houses, domes, zomes, and tree houses all have their thermal advantages, but they look a little weird next to that brick split-level or the wood-frame mansionette down the block. There are many less odd and less glamorous energy-efficient houses which look compatible with the other houses on the block. In fact, many of the best energy-saving ideas are incorporated into a new model building code devised by various governmental entities and adopted by ASHRAE, the American Society of Heating, Refrigeration and Electrical Engineers.

The ASHRAE code is complicated, but it does prescribe new standards for insulation, lighting, window design, and other aspects of home construction that will reduce energy costs. These will of course influence your house design. The model building code is now being evaluated by a number of organizations around the country, and will doubtless soon be incorporated into mandatory building codes. Some states have already made some of the ASHRAE recommendations a part of state building regulations. The energy standards in this model code will take the place, for example, of many inadequate insulation requirements that now exist.

If you want further information on the model code, or if you would like to incorporate some of its suggestions into your own new (or old) house, write to the National Conference of States on Building Codes and

Standards, 1970 Chain Bridge Rd., McLean, Virginia 22101.

The Importance of Insulation in Your New Energy-Saving House

Will changing the insulation requirements for new houses really make much of a difference? The best answer comes from Arkansas, where a unique plan of insulation and energy-saving construction techniques has gotten quite a bit of attention. The Arkansas Plan, as it is called, was developed in the 1960s by some advanced thermal thinkers, such as Harry Tschumi. You wouldn't think that insulation could demand much serious thought, but it took a lot of vision and experimentation to combat so-called conventional wisdom. Tschumi and others developed a whole new approach to tightening a house.

One of their discoveries, among many others, was the importance of thermal balance. A poorly insulated house does not waste fuel merely because of heat losses, but also because of *where* those losses occur. The poorly insulated house, Tschumi found, was often colder in one place—say, the floor—than in another, such as the top of the walls. To get the floor to a tolerable level of warmth, people advance the thermostat until the general temperatures throughout the rest of the house are higher than necessary. They use the thermostat not for its overall temperature-setting effect, but as a kind of troubleshooter, to remove cold air from specific areas. This troubleshooting technique causes high fuel bills, and is unnecessary in a house where the average temperatures are nearly the same at both floor level and ceiling level. One of the main purposes of insulation, Tschumi says, is to correct any thermal imbalance.

Tschumi developed the original specifications for well-insulated houses that use the heat pump. His ideas came off the drawing board when he teamed up with the late Les Blades, of Arkansas Power and Light, in 1961. They did some tests to determine the proper size of heating and cooling equipment needed in various types of residential construction, thus solving the common problem of oversized heaters and coolers. During the energy crisis of 1974, some of their experiments seemed to provide an answer to how to heat and cool with less fuel. Their plans, together with those of Frank Holtzclaw, a design analyst for HUD in Arkansas, created the home design that is now called the Arkansas Plan.

The first interesting thing about the Arkansas design is that it calls for 2"×6" studs instead of the conventional 2×4s that are the trademark of American construction. With 2"×6" studs, you can stuff six inches of insulation into the 5½-inch walls. The idea is to use more insulation and less wood, because the 2"×6" studs can be spaced farther apart than the 2×4s. Construction costs for the two types of building are almost equal, even though the Arkansas Plan house calls for: 12 inches of insulation in the ceiling; double-pane insulated windows or single windows with storm panels; 6 inches of insulation over the crawl space; a layer of foam insulation all around the perimeter of the slab; a plastic vapor barrier that completely covers the inside of every wall surface in the house; a sole-plate set into a full bed of caulking; electric wires that are strung so as not to impede the blanket of insulation; and heat pumps for heating and cooling.

The reported energy savings from the Arkansas Plan design, equal in cost to conventional construction, are substantial. According to Arkansas Power and Light, two 1,500-square-foot homes built under the plan in Benton, Arkansas, averaged $10.74 a month for heating and cooling costs during 1975. The utility said that monthly costs would have been at least $30.85 for a standard house. The two energy-conserving homes had saved about 65 percent on their heating and cooling. Arkansas electricity only cost about 2 cents per KWH at the time these figures were published, but the percentage of savings would remain the same, whatever the price of electricity. In fact, when electricity costs more, the Arkansas Plan reportedly returns much greater dividends.

Since that optimistic report, more than two hundred

new homes have been built in a contiguous area with Arkansas Plan specifications. Other builders, enamored of the idea that you could heat and cool a home for about $100 a year, have started putting up Arkansas Plan homes all across the nation. They have confirmed the original designers' contention that this house is no more expensive to build than a conventional house. The savings on wood make up for the increased cost for insulation, and in some cases construction time has even been reduced by the Arkansas design. Some of the newer models have been metered for an exact calculation of their energy consumption; the average annual heating and cooling costs for four such homes was 10.9 cents per square foot, as compared to 38.2 cents per square foot (at Arkansas electric rates) for comparable homes that conform to Minimum Property Standards.

The Arkansas Plan does not have to be accepted as doctrine. Many of its better features can be successfully adapted to a wide variety of building styles. The amounts of insulation it recommends might be altered to conform to climate conditions in other areas of the country. Arkansas requires an insulation value of R-38 for ceilings and R-19 for the walls. You couldn't pile that much extra insulation into an existing home, and even if you are building from the ground up, you might decide that such high levels are impractical in the design you have chosen.

But the ideas behind the Arkansas Plan are valid everywhere. To get major energy returns, you need much more insulation than is put into most new homes, even today. You also need better solutions to the problems of air infiltration, such as plastic sheets that cover the wall from top to bottom before the paneling is tacked on. Your caulking should be put down during construction, and the studs of the soleplate should be imbedded in it. You need to give more thought to the kinds of appliances, furnaces, and lighting devices that are to be installed in your new house.

You might give careful consideration to the heat pump, which was a major factor in the apparent success of the Arkansas Plan.

An Energy-Saving House in Your Future?

If you are interested in finding out more about the Arkansas Plan, the Owens-Corning Fiberglass Company puts out several booklets describing the project. You can get these by writing the company at Fiberglass Tower, Toledo, Ohio 43659.

An Energy-Saving House Heated and Cooled by Water

Another interesting energy-saving housing concept is ACES, or Annual Cycle Energy System. The ACES house gets its heating and cooling power from a large tank of water, three thousand cubic feet or more, that is placed in the basement. A heat pump is used, but not in the usual way. Instead of extracting heat from the air, which is what most heat pumps do, the ACES heat pump takes heat from the water in the basement tank. Over the winter, as more heat is removed from the water, it gradually turns into a huge chunk of ice. In the summer, the heat pump reverses itself, and melts the ice pack, providing a very efficient cooling system. The cost of an ACES heat-pump system would be $1,700 above the cost of a conventional central heating and cooling setup, but it will save at least $400 a year in energy bills.

Several private developers have adopted the ACES idea, and a company called Ellis Homes is using ACES in a development in Richmond, Virginia. For further information, you can contact ERDA, Office of Public Affairs, Washington, D.C. 20545.

The Almost-Solar and Fully-Solar House

It is of course much easier to consider solar energy for a house that hasn't been built.

There are many ways to go solar, but most successful solar houses rely first on *passive* technology; that is, they use window placement and careful site planning to maximize the natural contribution that the sun can make, and worry about the collectors and storage tanks later.

Some very interesting designs for solar houses are discussed in *Popular Science* magazine. In the December 1976 issue, a group of New Mexico architects were quoted as they discussed solar houses. One of the most interesting houses was the Sunscoop, which is built of adobe and uses natural sunlight to provide much of the heat. A huge glass area faces south, so heat can be retained; the Sunscoop house uses a Franklin Stove as a backup heater in the winter. Other solar houses discussed use huge water tanks, or use rock piles for heat storage; and some of the houses are built into the side of a hill. Another article, in the July 1976 issue of the magazine, and written by Richard Stepler, describes the general characteristics of the construction of most solar houses:

(1) They are designed from the ground up.
(2) They have no windows on the north side.
(3) They have extra insulation in walls and attic.
(4) The earth is back-filled along the north, east, or west walls, providing natural insulation.
(5) They use insulating drapes or shutters at night.

These *pre*solar modifications should be familiar by now. They are all things that will save you energy in your own home—solar or no solar.

Solar architects abound, and if you are interested, you will have no trouble finding your own. Solar house plans—incorporating collectors—have already been drawn up. One such plan, devised by an architect named Malcolm Wells, uses Thomasson solar collectors, a popular collector design. *Popular Science* reports on one family that built such a house. The family spends only $60 a year for backup oil heat, whereas in a regular house the fuel oil would cost them $750. The extra solar equipment in the house cost $5,500, in a total building price of $60,000; but the fuel savings are impressive. The plans for this house are available from Edmund Scientific Co., 700 Edscorp Building, Barrington, New Jersey 08007. The complete book of specifications and drawings costs $24.95. A license from the original designers to build such a house cost $40.

One of the most interesting—and most recently developed—solar-type houses is being used in a project called Minimum Energy Dwelling (MED), developed by the Southern California Gas Company, the Mission Viejo Construction Company, and the Energy Research and Development Administration (ERDA). The project houses use solar collectors for heating water and for heating and cooling the houses. It is also one of the first residential tests for a solar air conditioner. Two identical MED houses have been constructed in California. One will be occupied by a family, the other monitored by various devices that will measure its performance and its thermal characteristics. If you are interested in the design of the MED houses, or want information about how the project is going, you can write to Southern California Gas, MED Project, Research Department, P.O. Box 3249, Terminal Annex, Los Angeles, California 90051.

An Energy-Efficient Residence, or EER

A new experimental home called the Energy-Efficient Residence, designed by the National Association of Home Builders in cooperation with the Department of Housing and Urban Development (HUD), includes so many energy-saving features that describing it is a good way to list all the major possibilities. The EER residence draws on ideas from the Arkansas Plan, from some of the window strategies we have mentioned, from the "old-timer" factor, and from most of the products in the energy store—to create what is thought to be the ultimate in energy conservation! The EER house is the conservation version of the fully loaded car. The supporters of the project estimate that the extra cost of constructing such a house and buying the energy-efficient appliances will be paid back within five to seven years. At a conservative electric rate of 3.5 cents a KWH, this house is expected to save $630 a year over a typical electrically heated home with the standard amount of insulation.

A look down the list, and you'll know what to put in your EER house:

Design Features

(1) The house is compact, and rectangular, which minimizes heating and cooling loads.
(2) A storage room provides a windbreak on the end wall.
(3) An "airlock" entrance keeps heated air from escaping.
(4) A family retreat—a survival room—can be closed off, and heated or cooled separately.
(5) South-facing windows provide solar heat in winter.
(6) An ample roof overhang shades those windows in summer.
(7) Deciduous trees provide shading on the south side of the house.
(8) The windows that face north are small and are limited in size to 8 percent of the floor area.

Construction Features

(1) Plastic film barriers are stretched beneath the slab and behind the concrete-block walls.
(2) Exposed walls use R-19 insulation.
(3) Below-grade walls use R-11 insulation.
(4) There is a plastic-foam insulation, 2 inches thick, set around the perimeter of the slab.
(5) A one-inch fiberglass sill sealer is placed between the foundation and the sill plate.
(6) All utility entrances are sealed with heavy caulk.
(7) There are storm windows in the basement.
(8) A continuous plastic-film-vapor barrier is used behind the drywall.
(9) Electric outlets are mounted on the surface of walls and wiring is done in the floor. This way, the vapor barrier and insulation are not weakened with holes or gaps.
(10) The entrance door is closed magnetically.
(11) Windows are double-insulated, and storm

windows are added on top of that—which provides a triple-glazing effect.

(12) Insulating drapes are used.

(13) The window area is reduced to 8 percent of the floor area, except on the south side.

(14) Twelve-inch insulation batts are used in the attic ceiling. This gives an R-value of 38.

(15) Ceiling and gables are well ventilated.

(16) Lighting fixtures do not penetrate the ceiling. (Such recessed fixtures would create a major heat loss.)

Heating, Cooling, and Appliances

(1) There is a simplified duct system and the inside registers are kept low. (Where you put your heating or cooling registers is very important. Since hot air rises, the heating ducts should be located at floor levels; and since cold air falls, cold-air ducts should be placed high up on the walls. Most houses use the same ducts for both heating and cooling—a practice that results in higher energy bills.)

(2) A reduced-capacity heat pump is used. The compressor is installed indoors.

(3) A heat-recovery device provides free hot water when the heat pump is in operation.

(4) There are manually controlled, zoned heaters in the bathroom.

(5) The fireplace is specially designed for heat circulation. It has a glass-door enclosure.

(6) The hot-water heater is heavily insulated, and the thermostat is set at 120° F.

(7) Water pipes are insulated.

(8) Water-saver heads are provided on faucets and showers.

(9) A high-efficiency refrigerator takes the place of the standard variety.

(10) There is an energy-saving dishwasher with Air-Dry setting; a well-insulated electric range; a front-loading clothes washer that uses less water; fluorescent lighting wherever appropriate; bathroom vents with

dampers, to close down the exhaust when necessary; and a microwave oven.

A Traditional House with Some Energy-Saving Features

Even if you don't decide to build or buy a complete energy-saving house, you can still benefit by making slight alterations in a traditional house plan or house. Many such plans, which are sold for a small cost by the government and other sources, were drawn up before energy cost was much of a factor in home design; but they can be modified, in some cases, to create fuel economy. The National Bureau of Standards book *Window Design Strategies to Conserve Energy*, available from the U.S. Government Printing Office (Washington, D.C. 20402) for $3.75, contains an analysis of many such small modifications and how they might affect fuel costs. Here are some of the important ones from that booklet and from other sources:

Windows

The business of sunlight is a complicated one that has intrigued builders since the ancient Mayans. The heat value we get from sunlight depends upon the angle at which the light strikes a window, and those angles change with the latitudes and with the seasons. As a National Bureau of Standards study on windows says, "The further north a site is, the greater are the seasonal northerly and southerly shifts of sunrises and sunsets, the lower the arc of the sun across the sky." We can't possibly deal with all the variations here, but make sure your architect is thinking about sunlight when he or she designs your house (or has thought about it, if the house is already built).

A simple change in the placement of a couple of windows can have tremendous effect on the fuel consumption of a house. In the window study mentioned above, an experiment was conducted to see what would happen when 80 square feet of window was moved from the north to the south side of a house.

An Energy-Saving House in Your Future? 239

The house in question had three hundred square feet of window area, one hundred feet on the south and north sides, fifty feet on the east and west sides. By merely shifting the eighty feet, the heating requirement of the house was reduced from 92 million BTUs to 83 million BTUs, for a saving of 9 million BTUs. That could represent a saving of $45 on oil heat, or $140 on electric heat, based on an oil price of 50 cents per gallon and electricity at 5 cents per KWH.

The kind of windows you use is very important, too. By switching from single- to double-glazing, or to storm windows, you save money even if the house only contains a minimum of window area. If you are designing a house, consider tilting the windows slightly toward the ground. This will reduce heat gain in the summer without affecting heat gain in the winter—when you need it. The tilt increases the angle at which the sun strikes the window, reflecting more of the heat-producing rays. Since the sun is higher in the summer sky than in the winter sky, the tilt will be much more effective in summer than in winter.

A study done by the Edison Electric Institute, reported in the National Bureau of Standards booklet, reports that a typical all-electric ranch house will save 3,266 KWH of electricity with insulated glass instead of single-pane glass for a climate area like that of Indianapolis. This amounts to at least $130 a year. There are many kinds of insulated glass: glass with thermal barriers, double- and triple-glazings, glass with reflective film or even heat-absorption film, etc. You can check with local outlets to decide what is best for your area.

Everybody who has studied residential energy concludes that sliding glass doors are a thermal disaster. If you must use them, make sure you have double-glazed doors and that the sliding edges are properly weather-stripped.

Overhang

It is important to shade the southern exposure of a building. You can do it with deciduous trees—which lose their leaves when you want the sun, and get leaves back when you want shade—or with various kinds of louvered awnings. Or you can make sure you have a good roof overhang on your home.

A proper overhang is a matter for careful calculation. You want to maximize the amount of sunlight that falls on the house in the winter, and minimize it in the summer. The size and angle of your overhang depends on the angle of the sun as it goes through its various seasonal cycles at your latitude. You can now get a precise reading of such angles through valuable charts called Bennett Sun Angle Charts; they are available for every 2° of latitude from 24° through 52°. When ordering, specify your latitude and send $2 for the charts, plus 50 cents for postage and handling, to Robert Thomas Bennett, 6 Snowden Rd., Bala Cynwyd, Pennsylvania 19004.

Furnace

The main thing is size. Most houses have furnaces that could be much smaller, and therefore much more economical to operate. Contractors and architects have formulas to estimate the size of air conditioners and furnaces, but they then usually add on a generous "fudge" factor. You may not want to pay the fuel bill for such an overly cautious calculation. Find out the formula that your builder uses to come up with furnace or air-conditioner requirements, and if it sounds like too much "fudge," challenge it.

If you use electricity, remember that heat pumps may cost you less than half as much to operate as electric-resistance heaters. And if you do install a heat pump, by all means get the heat-recovery unit for free hot water. For electric homes, a heat pump and heat-recovery unit is a combination that would be hard to beat.

Also consider zoned heating. Most studies conclude that zoned heat is much cheaper than central heating or air conditioning. Some of the newer zoned systems contain dampers or valves in the ductwork that can be

An Energy-Saving House in Your Future?

turned off or on to easily control temperatures in various parts of the house.

If you decide on a central heating system, see if you can install a combination oil- and wood-burner. That way, you will always have some fuel, even in an oil embargo or shortage. Locate your heater in a central place in order to minimize duct runs. Insulate your ducts.

Appliances

Use life-cycle costing. Don't buy anything with a pilot light. Avoid excessive lighting. Avoid recessed lights. Get exhaust fans for the kitchen or bathroom that have dampers or flues, so that they can be turned off to preserve heated or cooled air. If you are using room air conditioners, consider installing them into the walls. That will free the windows and may encourage you to use the windows more often.

Carports and Garages

Carports, garages, vestibules, mud rooms, and other unheated or uncooled spaces can be used as effective windbreaks. If you design a house, put the carport on the north side of the house. The solar greenhouse that we discussed earlier (Section 7) can also be made a part of an original home design, and at a cheaper cost than if you add it to an existing building.

Where to Put Things

If you design a house—even a traditional one—you have a chance to locate your appliances in the right places. It's a costly process to move things like water heaters in existing houses, but it doesn't cost anything extra to get it right the first time. Here are some suggestions:

(1) The water heater, as we have mentioned, should be located as close to the sinks and faucets as possible. Many builders put all the plumbing fixtures in a central core, to minimize the expense of extra

pipe. The same technique will help minimize energy bills.

(2) The refrigerator should be kept away from the stove and away from a warm outside wall.

(3) The furnace should be in a warm part of the house, not someplace out in the cold.

(4) The air conditioner will work more economically if it is shaded and placed on the cold, north side of the house.

(5) The fireplace, if you are building one, might be less of an energy loser if it is placed in the middle of a room instead of on the outside wall. The infrared scanner used in the Massachusetts energy project detected major heat losses through the fireplaces and chimneys that were built on an outside wall of a house. The infrared findings support the wisdom of old-timers—who had their fireplaces and chimneys rising through the middle of their dwellings.

(6) The survival room, if you have one, should be located in a place where it can be closed off and heated or cooled separately.

(7) The living area should be built near the kitchen in sections of the country where cold predominates, and away from the kitchen in places where hot weather is the rule.

Adding a Room to an Existing House

Even if you aren't building a new house, you may discover that adding to your existing one can actually reduce cooling and heating bills. A solar greenhouse, of course, can have that effect, but so can other types of traditional construction. The benefits from such a project depend on what kind of house you have. If your house is box-like—a square or rectangle—you probably don't stand to gain from an energy-saving addition. But if your house is U-shaped, then you might.

Many Florida homes are built on the U pattern, with an open porch in the middle of the U. While such a porch makes an inviting and attractive selling point, in reality it is rarely useful for more than a few weeks out of the year. In the summer, this open porch is too hot, especially if the roof is made of thin, uninsulated materials. And, when sliding glass doors connect such a porch to the main house, it can add appreciably to the air-conditioning bills for the rest of the house. In the winter, the porch room gets too cold, and valuable furnace heat is cooled off on the surface of the sliding glass doors.

Many people have enclosed such porches, to make their houses into a rectangle. They have had to spend some money for the extra wall, and for insulation for the porch roof, but the results have often been most gratifying. In one such Florida home, the occupants report that their air conditioner used to run constantly with the U shape, but with the elimination of the porch they now use the air conditioner only about 25 percent of the time. Their winter heating costs have been reduced, too. This is a good example of energy-saving potential to be realized by structural modifications.

SECTION 14

Conclusion: What If Everybody . . . ?

A recent television special on energy raised the shudder-producing prospect that the country will some day be staring at the bottom of the barrel. The program emphasized increased production as the way to keep the barrel full. Each segment of the special began with the promise of one fuel or another—oil, nuclear, coal, natural gas—but ended with so many doubts and reservations that the viewer was kept bobbing between optimism and despair, and remained confused right up to the last commercial break. The final, indelible impression was that nothing will work well, and that there isn't much that anybody can do to avert an energy crisis.

The program didn't waste many accolades on the potential of energy conservation, which was not even brought up until viewers were about two hours into the show. Insulation was mentioned; so was turning down the thermostat. But the idea that the public could really help solve the energy problem was dispelled by the people-on-the-street interviewed on the show. Many of them didn't know that the United States must now import much of its oil. The prevailing impression one received from listening to these people was that Americans would continue to overheat, overcool, overcook, overlight, and overdrive themselves until the last drop of oil is wrung from the last wellhead in Saudi Arabia. Americans will choke on oil, the interviews implied, just as the Greeks and Romans choked on decadence.

As it stands now, the United States contains about ercent of the world's population, and uses some-

Conclusion: What If Everybody . . . ?

where over 30 percent of the world's energy. Our consumption equals the *equivalent* of 35,000,000 barrels of oil a day; and included in that amount is the 18,000,000 barrels of actual oil. (We get the rest of our power from nuclear, hydroelectric, and other sources.) Of the 18,000,000 barrels a day, 6,000,000 have to be brought in from other countries.

The Federal Energy Administration gives some estimates of how that demand could be affected, if energy saving ever got popular:

(1) If every household in the United States lowered its winter thermostat setting an average of 6°, the country would save 570,000 barrels of oil per day. That's enough to heat over 9,000,000 homes during a winter season.

(2) If every house raised its summer thermostat by 6°, the country would save the equivalent of 36,000,000,000 kilowatt hours of electricity, or 2 percent of the nation's yearly electricity consumption.

(3) If every house were caulked and weather-stripped, about 580,000 barrels of home heating fuel would be saved for every winter day. If the 18,000,000 single-family homes without storm windows were given such windows, the national fuel demand would drop about 200,000 barrels each day of the winter. Universal attic insulation would save about 400,000 barrels of heating oil a day.

The Consolidated Edison of New York utility has brought these calculations down to the home level. It takes 19 gallons of oil, or about half a barrel, to make enough electricity to light one 100-watt light bulb for six hours a day for one year. The average family, Con Ed says, uses one barrel of oil for lights, two for the refrigerator, three for the stove, three for the air conditioner, ten for hot water, plus four gallons for the toaster and seven pints for the electric toothbrushes. You can figure out your own oil consumption by remembering that one barrel of oil—or about 42 gallons —will generate about 500 kilowatt hours of electricity. Thirty hours of watching television each week uses up a barrel of oil a year.

If every family could reduce residential consumption of oil by 25 percent, the amount saved would replace one-third of our dependence on foreign oil. As we saw in the Seattle case, such a reduction would also change the entire investment future of utility companies, and eliminate many costly and controversial new plants. The increased awareness that would accompany such a widespread *residential* reduction would force more *commercial* and *industrial* enterprises to do their part.

But the remaining question is whether energy saving will get popular. Conservation is viewed cynically by many energy policymakers, even the ones who release the "What if?" statistics. They think energy conservation is good as an afterthought to increased production, good as sermonizing material; but they doubt the strength, ability, and intelligence of the American people to carry it out. Those doubts are dispelled—in our opinion—by the results of careful conservation planning in places like Seattle and Los Angeles. We think conservation can be the first—and not the last—solution to energy problems.

We give two reasons. The first is the positive public response that has been achieved in places where conservation was an immediate necessity. People have understood conservation when it was presented as an immediate economic choice, and not as a matter of saving the dinosaur's legacy for future generations. The second is that conservation is in people's own self-interest. Many among us have valid reasons to distrust the patriotic appeal to save energy, since campaigns that start out to protect the nation's resources often end up emptying the nation's citizen's pocketbooks. But there are personal benefits to conservation that cannot help but be attractive. When those benefits are accepted and believed in, when they are understood as more than advertising gimmicks by insulation companies, then people will go after them. Collective rewards are sure to follow from the personal.

Acknowledgments

We would like to thank the following:

Janet Michaud, for diligent research; Helen Pardee, for around-the-clock typing and editing services; Evan Powell, one of the best energy journalists around, for advice and encouragement; Bob and Polly Knox and Pat Jaffe, for the use of their premises; Ed Moran and *Popular Science,* for a lot of good leads and useful ideas; Bethany Widener, of Sen. James Abourezk's staff, for research help; Loretta DeCamille, of Con Ed and Pat Logie of Seattle City Light and Power, for doing more than just sending information; and Susan Wolf Rothchild and Joyce Tenney for commentary and editing (and prodding).

Our thanks also go to the people at the National Bureau of Standards, ERDA, Princeton University, Johns Hopkins, the Davis California energy project, the Environmental Action Foundation, the Association of Home Appliance Manufacturers, the Owens-Corning Company, the Florida Solar Energy Center; and to Pat Baker in the Massachusetts Energy Office, Peter Hincks of Vermont Castings Company, William Hillner of Daystar Corporation, G. J. Burrer of Inframetrics, Inc., Vincent D. DiCara in the State of Maine Office of Energy Resources, and Harvey Haines, owner of Haines Refrigeration and Appliances, for his many practical suggestions.